Self-Assembling Amphiphilic Systems

Phase Transitions and Critical Phenomena

Edited by

C. Domb

Department of Physics,
Bar-Ilan University, Ramat-Gan, Israel

and

J. L. Lebowitz

Department of Mathematics and Physics,
Rutgers University, New Brunswick,
New Jersey, USA

Volume 16
Self-Assembling Amphiphilic Systems

Gerhard Gompper

Sektion Physik der Ludwig-Maximilians
Universität München
80333 München, Germany

and

Michael Schick

Department of Physics,
University of Washington,
Seattle, WA 98195, USA

ACADEMIC PRESS

Harcourt Brace & Company, Publishers

London   San Diego   New York
Boston   Sydney   Tokyo   Toronto

ACADEMIC PRESS LIMITED
24–28 Oval Road
London NW1 7DX

U.S. Edition published by
ACADEMIC PRESS INC.
San Diego, CA 92101

A catalogue record for this book is available from the British Library
ISBN 0-12-220316-X

Typeset at Alden Press, Oxford and Northampton, Great Britain
Printed in Great Britain by TJ Press Ltd, Padstow, Cornwall

Contents

Self-Assembling Amphiphilic Systems

G. Gompper and M. Schick

General Preface

This series of publications was first planned by Domb and Green in 1970. During the previous decade the research literature on phase transitions and critical phenomena had grown rapidly and, because of the interdisciplinary nature of the field, it was scattered among physical, chemical, mathematical and other journals. Much of this literature was of ephemeral value, and was rapidly rendered obsolete. However, a body of established results had accumulated and the aim was to produce articles that would present a coherent account of all that was definitely known about phase transitions and critical phenomena, and that could serve as a standard reference, particularly for graduate students.

During the early 1970s the renormalization group burst dramatically into the field, accompanied by an unprecedented growth in the research literature. Volume 6 of the series, published in 1976, attempted to deal with this new literature, maintaining the same principles as had guided the publication of previous volumes. The number of research publications has continued to grow steadily, and because of the great progress in explaining the properties of simple models, it has been possible to tackle more sophisticated models which would previously have been considered intractable. The ideas and techniques of critical phenomena have found new areas of application.

After a break of a few years following the death of Mel Green, the series continued under the editorship of Domb and Lebowitz, Volumes 7 and 8 appearing in 1983, Volume 9 in 1984, Volume 10 in 1986, Volume 11 in 1987, Volume 12 in 1988, Volume 13 in 1989, Volume 14 in 1991 and Volume 15 in 1992. The new volumes differed from the old in two new features. The average number of articles per volume was smaller, and articles were published as they were received without worrying too much about the uniformity of content of a particular volume. Both of these steps were designed to reduce the time lag between the receipt of the author's manuscript and its appearance in print.

The field of phase transitions and critical phenomena continues to be active in research, producing a steady stream of interesting and fruitful results. It is no longer an area of specialist interest, but has moved into a central place in condensed matter studies. The editors feel that there is ample scope for the series to continue, but the major aim will remain to provide review articles that can serve as standard references for research workers in the field, and for graduate students and others wishing to obtain reliable information on important recent developments.

CYRIL DOMB
JOEL LEBOWITZ

Preface to Volume 16

This volume is concerned with molecules with a hydrophilic (water-loving, polar or ionic) head, and a hydrophobic (water-hating, usually hydrocarbon) tail. In 1936 G. S. Hartley introduced the term *amphipath* to describe such molecules; this was later replaced by *amphiphiles* 'loving both', an expression which has gained wide acceptance. The subject may well be unfamiliar to research workers in the area of phase transitions and critical phenomena and we, therefore, devote a few paragraphs to an introduction to the terminology and history of the field.

The above-mentioned special characteristics of these molecules give rise to their self-assembling properties. For example, at an oil–water interface the molecules will line up to form a molecular layer with heads in the water and tails in the oil. As a consequence, the surface tension is reduced, an effect known for many years. However, the fact that the reduction can be very large for a very small concentration of some amphiphiles has not been well understood. Those amphiphiles which exhibit this property are often referred to as surfactants or surface active compounds.

Similarly when wholly immersed in water, the molecules again try to reduce the area of contact of the tails with water and can form spherical aggregates called micelles in which the heads are on the surface of the sphere and the tails point inward. Typical micelles are aggregates of 100 molecules.

A similar, but much larger object, in which a single bilayer encloses a water-filled cavity is called a vesicle.

When the concentration of amphiphile increases in an aqueous solution, phases can form in which the bilayers assemble themselves into lamellae, or into rods or spheres which then are arranged in a regular manner familiar from liquid crystals. Such systems, in which phase transitions are driven by changes in concentration, are called lyotropic.

Physical and colloid chemists have known for some time of the bewildering complexity of phase diagrams which arise in oil-water-amphiphiles mixtures. It is only in the past two decades that detailed theoretical explanations based on statistical mechanics have been forthcoming. The number of papers in the field has grown rapidly, and one of the useful contributions made by this volume is a critical survey of the widely scattered literature.

The discussions in this volume are somewhat different from those in previous volumes of the series. Emphasis is not on the detailed mathematical behavior near phase transitions or critical points, but rather on providing models with a minimum number of parameters which can account for the observed experimental complexity.

Three different approaches are considered. The first is the formulation and investigation of suitable models of ternary solutions. Even a mean field treatment poses challenging problems, and Monte Carlo simulations have helped to unravel the properties of the models. The second approach uses multivariable Landau order-parameter expansions whose coefficients are fitted empirically. Finally, the membrane approach starts with an established interface and considers the effect of fluctuations in it. Each approach has its strong and weak points which the authors outline, and the approaches are seen to be complementary.

Although the theory has not yet reached the stage of quantitative comparison with experiments, the qualitative agreement achieved is indeed impressive.

We are greatly indebted to the authors for doing such a splendid job.

CYRIL DOMB
JOEL LEBOWITZ

Contents of Volumes 1–15

*Out of print.

Contents of Volume 10

Contents of Volume 11

Contents of Volume 12

Contents of Volume 13

Contents of Volume 14

Contents of Volume 15

Self-Assembling Amphiphilic Systems

G. Gompper[1] and M. Schick[2]

[1]Sektion Physik der Ludwig-Maximilians Universität München, 80333 München, Germany. [2]Department of Physics, University of Washington, Seattle WA 98195, USA.

PHASE TRANSITIONS
VOLUME 16 ISBN 0-12-220316-X

We will unite the white rose and the red.
Smile, heaven, upon this fair conjunction,
That long have frowned upon their enmity!
Richard III, Act V: Sc. V

1 The phenomena and approaches to them

1.1 Introduction

Given that oil and water do not mix, it is astonishing that the addition of a small amount of amphiphile can spontaneously bring them together with the formation of a wealth of complex structures. These range from a disordered collection of simple oil-filled sacs to regular ordered arrays of cylindrical containers shielding the oil within from the water without. Their construction is driven by the properties of the amphiphile, a molecule with a polarizable or ionic head which prefers the highly polarizable water environment, and a hydrocarbon tail which prefers the oil. Hence "amphiphile", from the Greek "loving both". The energy of the amphiphile is lowest when it can find or create surfaces between oil and water at which it can adsorb, surfaces which then arrange themselves, and the other two components, into various structures.[†] It is because of this surface activity that such molecules are also referred to as surfactants. The simple sac-like aggregate is a micelle, a small, usually spherical structure in which the amphiphiles on the surface shield oil molecules inside from contact with water. The roles of oil and water are reversed in so-called "inverted" micelles. The tendency of amphiphile to organize the other two components finds its fullest expression in the lyotropic phases which occur when the concentration of amphiphile is large. Examples are the cubic phase, a cubic packing of normal or inverted micelles, the hexagonal phase, a close-packed array of long cylindrical micelles, and the lamellar phase which consists of sheets of amphiphile. Two of these phases are shown schematically in Fig. 1.1. If there is not enough amphiphile to overcome the disordering tendencies of temperature and to bring about a lyotropic phase, but more than enough to overcome the tendency of the oil–water interactions to induce phase separation, then a fluid phase can be formed in which the water and oil are solubilized and the ordering tendency of the amphiphile is still manifest. The result of this tendency may appear as a finite concentration of micelles within the fluid, denoted therefore a micellar solution, or as a disordered array of sheets of amphiphile

---

[†]General reviews are provided by Nelson *et al.* (1989), S.-H. Chen *et al.* (1992) and Gelbart *et al.* (1994). See also the conference proceedings of Meunier *et al.* (1987) and Lipowsky *et al.* (1992).

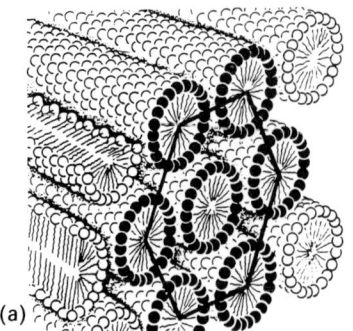 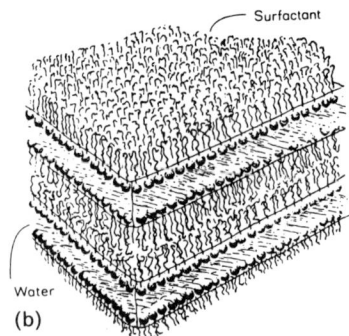

Fig 1.1 Schematic representation of (a) hexagonal and (b) lamellar phases. From Rosevear (1968).

separating coherent regions of water from coherent regions of oil. If the concentrations of the water and oil are not very different, then both coherent regions will span the system, and the fluid is said to be bicontinuous (Scriven 1976, 1977). Such a fluid, denoted a microemulsion, clearly needs a second length scale, the characteristic distance between amphiphilic sheets, to describe it in addition to the usual coherence length of the system. It is precisely this region of intermediate amphiphile concentrations, sufficient to produce a structured, or complex fluid, but not so much as to produce a lyotropic phase, which poses some of the most intriguing theoretical challenges, and will be of greatest interest to us.

Note that the microemulsion, like the lyotropic phases, is characterized by *extensive* amounts of internal interface which indicates that the free energy per unit area of such internal interfaces between oil and water is negative. If the amphiphile concentration is reduced so that oil and water again phase separate, we might expect that the interface between the bulk oil and water phases would be similar to the internal interfaces of the microemulsion. Although the free energy of the oil/water interface, which is just the interfacial tension, must be positive, the resemblance to the internal interface leads us to suspect that its value may be rather low. In fact the oil/water tension can be reduced by the presence of amphiphile to a value a thousand times smaller than that in its absence! Technological interest in such systems is based on this property, as well as on the self-assembly. An interesting example of the latter is in the use of the hexagonal phase to make a molecular sieve with holes the size of the cylindrical micelle, a size easily varied (Beck *et al.*, 1992).

Self-organization also occurs in the two-component system of water and amphiphile; we can imagine it by replacing the single layer of amphiphile, which forms between two regions of oil and water in the ternary mixture, with a double layer of amphiphile between two regions of water in the binary mixture. Lamellar phases composed of amphiphilic bilayers exist, and can be made with as little as one per cent by weight of amphiphile. With so little of this component, the lamellae are thousands of angstroms apart, a distance orders of magnitude larger than the size of the molecular constituents. At even lower concentrations, we can expect a fluid phase completely analogous to the microemulsion. Such a complex fluid does exist and is denoted the L_3, or sponge, phase. (For reviews, see Roux *et al.* (1992a) and Porte (1992).) This brief introduction to experimental observations on amphiphilic systems will be expanded upon below, as we wish to establish some of the phenomena which these systems exhibit and which, therefore, should be addressed by theory.

1.2 Some basic properties of oil–water–amphiphile mixtures

1.2.1 Phase behaviour

The phase behaviour of ternary mixtures of oil, water and amphiphile is interesting, displaying some features common to all ternary systems, and others characteristic of this particular system (Knickerbocker *et al.*, 1982; Firman *et al.*, 1985; Kahlweit *et al.*, 1985; Davis *et al.*, 1987). It is a common-place knowledge that oil and water do not mix, and adding a small amount of amphiphile to the system will not, in general, alter the two-phase coexistence. The added amphiphile will partition itself between the two phases. For the sake of argument, let us assume that the amphiphile is found predominantly in the water phase. By changing an external field, a third phase can be made to appear, one which contains more of the amphiphile and less of water and oil than the other two. Its density, therefore, will be intermediate between that of the water-rich and the oil-rich phases, so that it will be physically located between them. Hence it is denoted the middle phase. What the external field is depends on the type of amphiphile employed. For ionic amphiphiles, such as sodium-bis-ethylhexylsulfosuccinate (AOT), or sodium dodecyl sulfate (SDS), the chemical potential of a fourth component, salt, is often the external field, while for non-ionic amphiphiles, such as the homologous series of *n*-alkyl polyglycol ethers, C_iE_j,[†] it is the temperature. To be definite, let us assume that the system contains a non-ionic amphiphile. Then it is at low temperatures that the amphiphile is found predominantly in the water-rich phase which is in two-phase equilibrium with a nearly pure oil phase. The middle phase appears as the temperature is increased to some value T_L. On raising the temperature further, the amphiphile becomes increasingly more

[†]The notation is short for $CH_3(CH_2)_{i-1}(OCH_2CH_2)_jOH$.

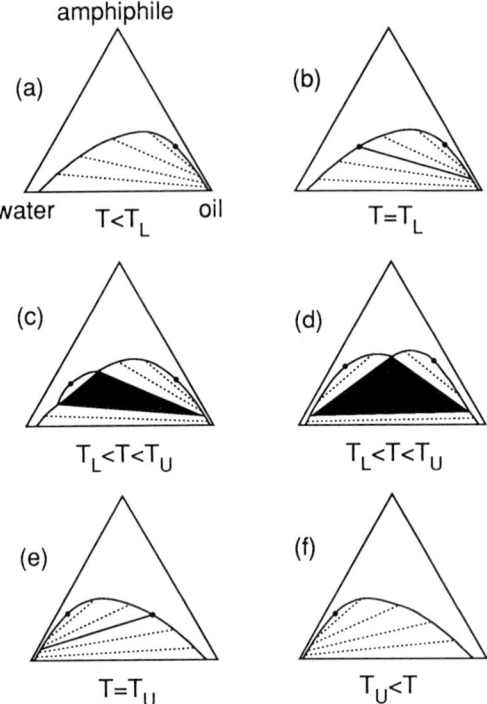

Fig. 1.2 Sequence of schematic phase diagrams, each at constant temperature, of a ternary system of water, oil and a non-ionic amphiphile. The temperature increases from (a) to (f). T_L and T_U are the lower and upper critical endpoint temperatures. Critical points are shown with a dot (●). For clarity, lamellar and other lyotropic phases are not shown.

soluble in oil and less in water, so that a point is reached at which the concentrations of oil and water in the middle phase are equal. The middle phase is said to be balanced. Increasing the temperature again, the composition of the middle phase becomes more and more similar to that of the oil-rich phase until, at some temperature T_U, these two phases become identical, and we are left again with only two-phase equilibrium. Now, however, the two phases are nearly pure water phase, and an oil-rich phase which contains most of the amphiphile. This sequence is shown schematically in Fig. 1.2 in a series of ternary concentration diagrams, each at a fixed temperature. The points at which the middle phase disappears by becoming critical with one of the phases in the presence of the third are critical end-points. The series of ternary diagrams are shown stacked upon one another in Fig. 1.3 to create the Gibbs phase prism. In this figure a line, cp_α, has been drawn through all the

non-ionic amphiphile (C)

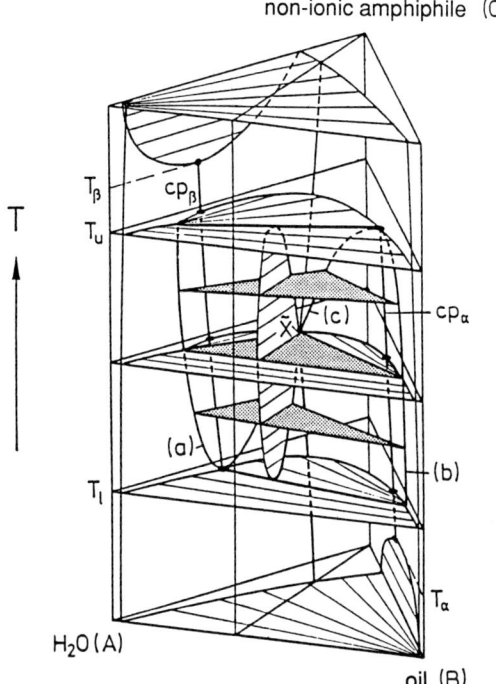

Fig. 1.3 Schematic phase prism of a ternary system oil, water, non-ionic amphiphile, with temperature as ordinate. The three-phase body appears at the lower critical end point T_L and disappears at the upper critical end point T_U. Also shown are the trajectories of the water, the oil and the microemulsion at three-phase coexistence. cp_α (cp_β) is the line of critical points between the oil-rich (water-rich) and middle phases. The dark triangles represent isothermal sections through the three-phase body. From Kahlweit *et al.* (1987).

points at which oil and middle phases become critical, and this line extends from the upper critical point at T_U down to the critical point of the binary system of oil and amphiphile. Similarly a line, cp_β, has been drawn through all the points at which water and middle phases become critical, and it extends from the lower critical endpoint at T_L to the inverted critical point of the binary water–amphiphile system.

In order to simplify the representation of the phase behaviour while retaining essential information, a cut through the phase prism at a fixed oil to water ratio is often shown. Such a cut at a ratio of oil to water 1:1 is shown schematically in Fig. 1.4. Due to the shape of the three-phase body, it is often referred to in the literature as "the fish". The highest part of the fish occurs at the upper critical endpoint temperature, T_U, and the lowest part at the lower

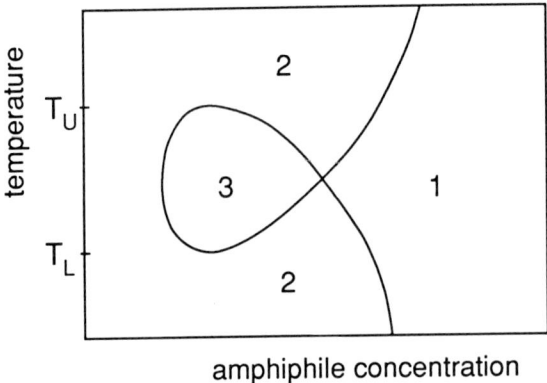

amphiphile concentration

Fig. 1.4 Cut through the Gibbs phase prism of Fig. 1.3 at a ratio of oil to water of 1 : 1 (schematic). Single-phase region is denoted by 1; regions of two- and three-phase coexistence are labelled 2 and 3.

critical endpoint, T_L. The head of the fish gives the concentration of the base of the three-phase triangle, at oil–water coexistence. The tail of the fish in this 1:1 cut gives the concentration of amphiphile in the middle phase, the concentration at the apex of the three-phase triangle of the balanced system. The extent of the fish, then, is just the height of the three-phase triangle. The concentration at the tail of the fish is a useful measure of the efficiency of the amphiphile as it gives the minimum concentration needed to produce, in this balanced system, a single phase in which the oil and water are solubilized. We shall refer to inefficient solubilizers as weak amphiphiles, and efficient ones as strong.

A great advantage of the non-ionic amphiphiles C_iE_j is that their strength can be changed systematically: weakened by decreasing both the number i of carbons in the hydrophobic tail and the number j of ethoxy groups, strengthened by increasing them. This can be seen in Fig. 1.5 where one sees several such diagrams for water–decane–C_iE_j systems. One sees that as one goes from C_8E_4 to C_6E_3 to C_4E_1, the tail of the fish moves towards larger amphiphile concentrations, hence the amphiphiles are increasingly weaker. One also sees, in going from C_6E_3 to C_4E_1 that the head also moves towards increasing amphiphile concentrations. There is reason to believe that if this discrete series of amphiphiles could be weakened further, the head would approach the tail, i.e. the height of the three-phase triangle would become smaller and the system would show a tricritical point at which the three-phase triangle shrinks to a point and all phases become critical simultaneously. In order to continuously decrease the amphiphile strength, a fourth component, such as formamide, can be added to the system (Kahlweit et al., 1991). The resulting phase diagrams are shown in Fig. 1.6 for four systems of water,

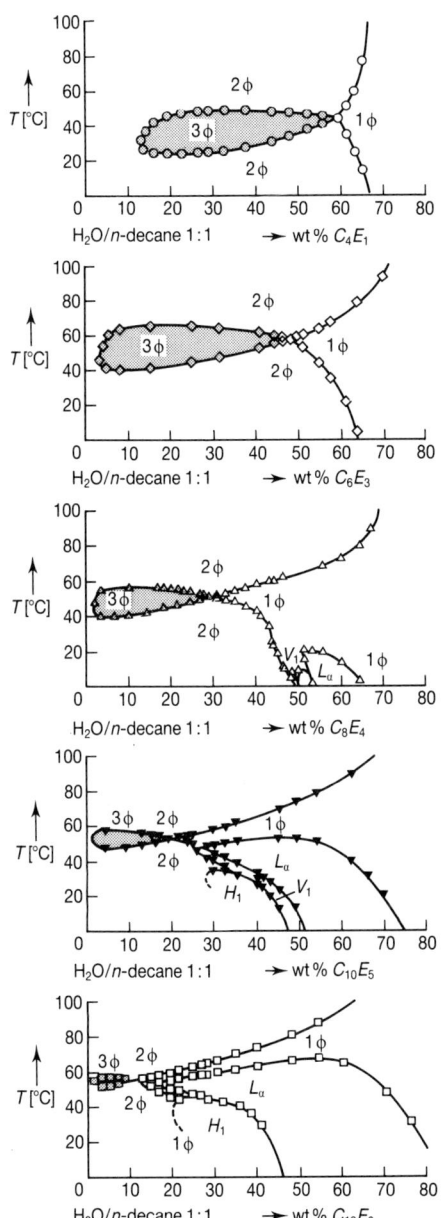

Fig. 1.5 Vertical sections through the phase prism at a ratio of oil and water 1:1, for the system water, decane and C_iE_j with increasing strength of the amphiphile. Reprinted with permission from M. Kahlweit, R. Strey and P. Firman, *J. Phys. Chem.* **90**, 671 (1986). Copyright 1986 American Chemical Society.

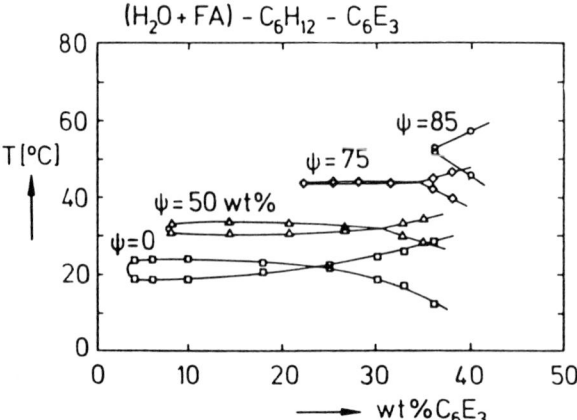

Fig. 1.6 Vertical section through a phase prism with the three-phase bodies of (water + formamide), cyclohexane, C_6E_3 at volume ratio (water + formamide)/oil = 1, with a formamide to water ratio ψ as parameter. From Kahlweit *et al.* (1991).

formamide, cyclohexane and C_6E_3, with different ratios ψ of formamide/(formamide + water). As the amphiphile becomes weaker, the extent of the three-phase triangle, both its height and its width (not shown), decreases and vanishes at the tricritical point. The two critical endpoints merge there, as can also be seen from the figure as the difference between the highest and lowest part of the fish vanishes.

The phase behaviour presented thus far, i.e. the two-phase to three-phase to two-phase progression as one field is varied, and the merging of the two critical endpoints at a tricritical point as another is changed, is just that expected for normal ternary mixtures (Furman *et al.*, 1977). The behaviour peculiar to the amphiphilic systems appears as the amphiphiles are made stronger. If we return to Fig. 1.5, we see that as one goes from C_8E_4 to $C_{10}E_5$ to $C_{12}E_6$, a lamellar phase, denoted L_α, appears and occupies much of the phase space in which the fluid phase was found. Also note from the last two diagrams that, at the temperature at which the tail of the fish occurs, one observes the progression with increasing amphiphile concentration noted earlier: oil–water coexistence, followed by the appearance of the middle phase, followed by the lamellar phase. As the strength of the amphiphile increases, the interval over which the middle phase exists decreases, and eventually its existence is completely pre-empted by the lamellar phase. Three-phase coexistence is now between oil, water and lamellar phases. Just before this occurs, one expects a four-phase point at which oil, water, middle and lamellar phases all coexist.

The appearance of lamellar phases is an obvious difference between the

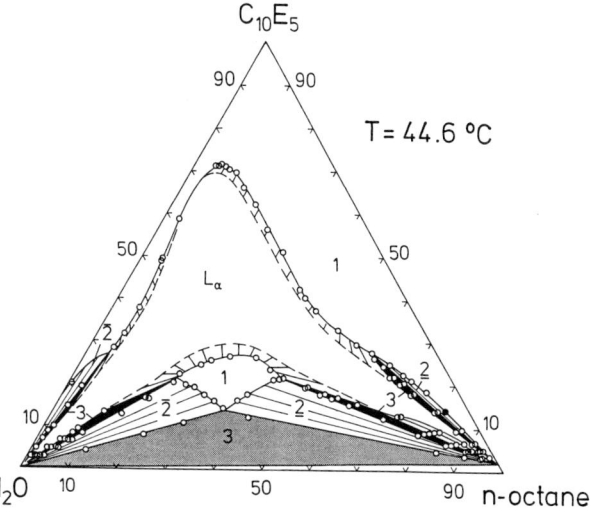

Fig. 1.7 Gibbs triangle of the system water, octane, $C_{10}E_5$, at temperature $T = 44.6\,^{\circ}C$. L_α denotes the lamellar phase. From Strey (1993).

phase diagrams of systems containing weak and strong amphiphiles. Their presence introduces another difference which we want to stress. Let us consider the phase diagram of a system containing a weak amphiphile at a temperature at which the middle phase is balanced. This would look like Fig. 1.2(d). We note that there is a continuous path from the water-rich phase to the oil-rich phase at that fixed temperature. An analogous phase diagram at fixed temperature is shown for a system with a strong amphiphile, $C_{10}E_5$, water and octane (Strey, 1993) in Fig. 1.7. We see here that the middle phase is isolated from the oil- and water-rich phases due to the presence of the lamellar phases. There still exists a continuous path from oil to water phases, but in order to traverse it, the temperature must be varied. This path is clearly visible in Fig. 1.8 which shows, *inter alia*, a cut through the phase diagram of the water–decane–$C_{10}E_5$ system at constant amphiphile concentration equal to that at the tail of the fish.

Phase behaviour of the binary water–amphiphile system is a bit simpler if only because there is one less chemical potential to be varied. The most prominent feature of the phase diagram is phase separation between water-rich and amphiphile-rich phases with a lower critical point. With increasing amphiphile concentration a variety of lyotropic phases are found. (For general reviews, see Tiddy (1980), Seddon (1990) and Hoffmann (1990).) The sequence in which they occur is usually hexagonal, cubic, lamellar, inverted hexagonal, but not all these phases are always present. In the

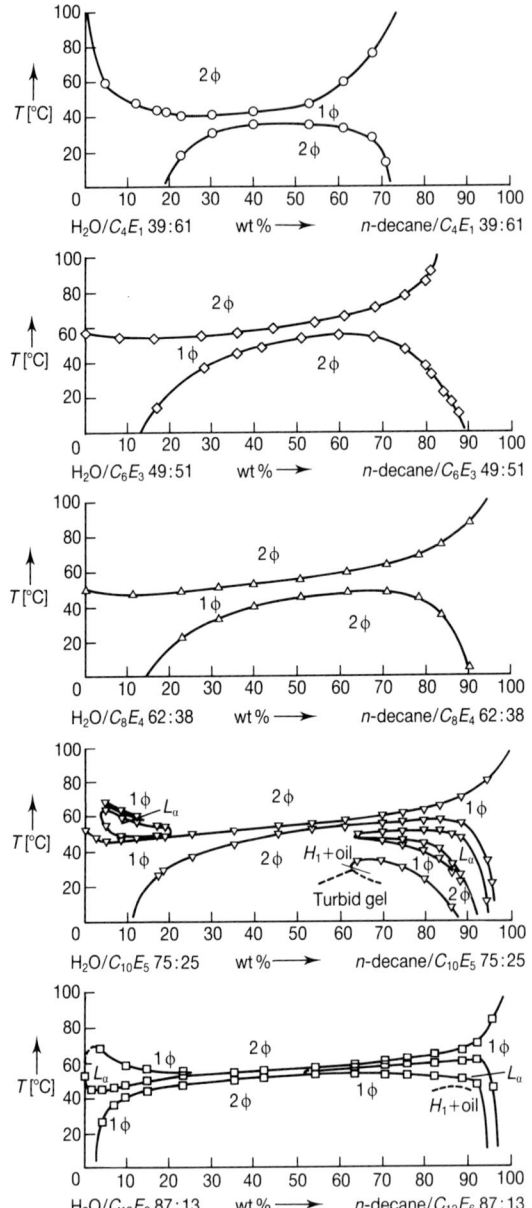

Fig. 1.8 Vertical sections through the phase prism at constant amphiphile concentration, for the system water, decane and C_iE_j with increasing strength of the amphiphile. From Kahlweit *et al.* (1986).

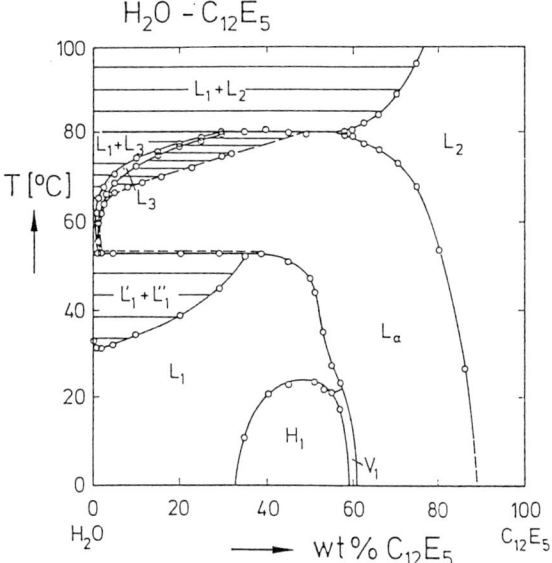

Fig. 1.9 Phase diagram of the water–$C_{12}E_5$ system. L_1, L_2 and L_3 denote isotropic liquid solutions, H_1 is a normal hexagonal phase, V_1 is a cubic liquid crystalline phase and L_α denotes a lamellar phase. From Strey *et al.* (1990).

inverted phase, there is so little water that it is inside the cylindrical micelles. A phase diagram for the system water and $C_{12}E_5$ is shown in Fig. 1.9 (Strey *et al.*, 1990). There is a pure water phase which is almost invisible at the left-hand edge. At low temperatures one sees the progression from fluid (L_1), to hexagonal (H_1), to cubic (V_1), to lamellar (L_α). At higher temperatures, unusual features appear. In a narrow temperature region around 60 °C, the lamellar phase swells to 98.9% water. Between about 54 °C and 80 °C, a new phase appears between the pure water and lamellar phases. This is the "sponge" phase (L_3). It extends from 70 to 99.5% water! It is very surprising that a phase which contains only 0.5% amphiphile can be distinct from the pure water phase itself.

1.2.2 *Interfacial properties*

Certainly one of the most spectacular effects in amphiphilic systems is the great reduction of the interfacial tensions observed in them. Fig. 1.10 shows the various tensions in the system water, salt, toluene, SDS and butanol. This system shows the two- to three- to two-phase progression as the concentration of salt is increased. At the critical endpoint at which the middle phase and the water-rich phase become identical, the interfacial tension between them,

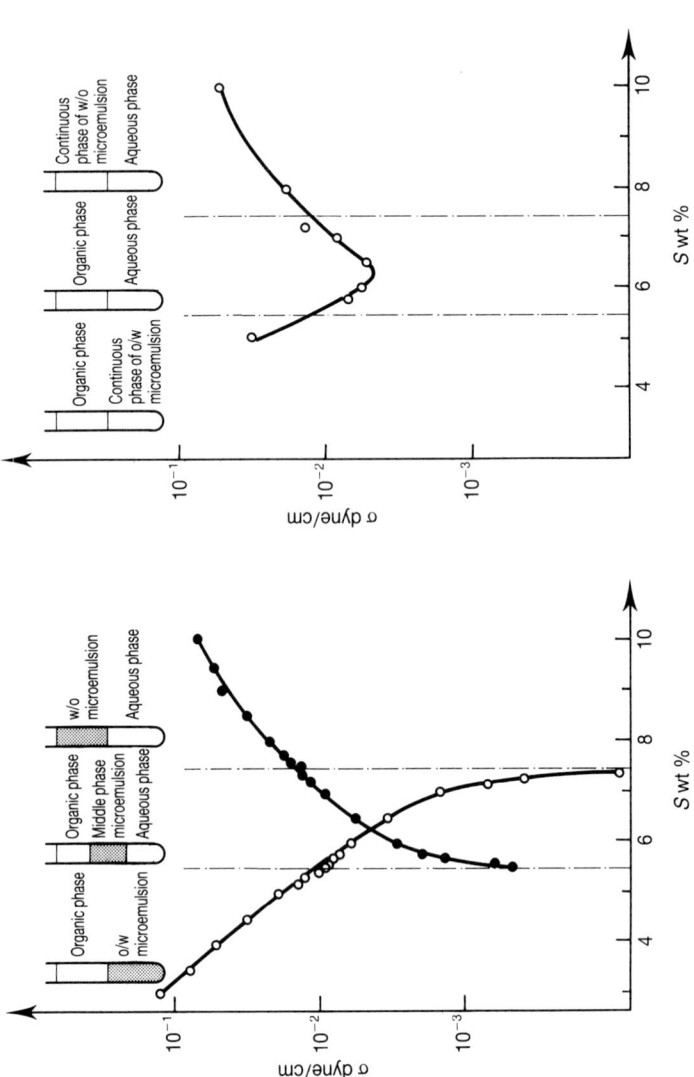

Fig. 1.10 The interfacial tensions of oil/water, oil/microemulsion and water/microemulsion interfaces, as a function of salinity S, in the system water, salt, toluene, SDS and butanol. From Pouchelon *et al.* (1980).

σ_{wm}, must vanish. Similarly, the interfacial tension between middle and oil-rich phase, σ_{om}, must vanish at the other critical endpoint. This accounts for the general shape of the curves in Fig. 1.10. The tension between the oil and water phases, σ_{ow}, which is only defined in the region of three-phase coexistence, does not vanish anywhere. It is lowest at some concentration of salt almost midway between the two endpoints. A constraint on how large it can be results from the inequality

$$\sigma_{ow} \leq \sigma_{om} + \sigma_{wm} \tag{1.1}$$

and this accounts for its general shape. (The necessity for (1.1) is clear: if it were violated, the system could reduce its interfacial free energy by inserting a macroscopically thick wetting layer of middle phase between oil and water, thus restoring the equality. This has already been clearly stated by Gibbs (1878).) Note the scale of the measured tensions, and compare them to normal oil/water tensions in the absence of amphiphile which are about 50 dynes/cm. Thus, the oil/water tension has been reduced by three orders of magnitude! One can immediately speculate on the reason for this reduction. If the two critical endpoints are close to one another, then the vanishing of one of the tensions at each endpoint and the constraint (1.1) enforce a small oil/water tension. This is certainly the case for weak amphiphiles driven close to a tricritical point. It is also true that the critical endpoints are indeed close to one another in systems of strong amphiphiles. This can be seen in Fig. 1.5 where the temperature difference between upper and lower critical endpoints in the system water, decane, $C_{12}E_6$ is quite small. It is tempting to believe that these strong systems are also close to a tricritical point, a point of view held for some time. We will argue below, however, that this is *not* the case for strong amphiphiles.

The other interfacial property which will be of interest to us is the following: at three-phase coexistence, a middle phase containing a weak amphiphile wets the interface between the oil- and water-rich phases, while one containing a strong amphiphile does not (Seeto et al., 1983; Aratono and Kahlweit, 1991). We would like to know the reason, if any, for this correlation, and what it tells us. Further, given a system in which the middle phase does not wet the oil/water interface, one expects from general arguments (Cahn, 1977; Rowlinson and Widom, 1982) that a wetting transition will occur as the system is driven to a tricritical point, by weakening the amphiphile, or to either of its critical endpoints, by varying the temperature. Beyond this transition, the middle phase will wet the oil/water interface. Such transitions have been observed (Robert and Jeng, 1988; L.-J. Chen et al., 1990; Smith and Covatch, 1990; Aratono and Kahlweit, 1991; Schubert and Strey, 1991), and one would like to predict the order of the transitions, and their approximate location.

1.2.3 Structure

There are several experimental methods which can be employed to determine that the middle phase does exhibit considerable structure, and to explore its nature. Perhaps the simplest, conceptually, are X-ray and small angle neutron scattering experiments. (For a review, see Auvray (1993).) The water–water structure function, shown in Fig. 1.11, shows a peak at non-zero wavevector q indicating that there is structure on a scale of $2\pi/q$. For this particular system water, salt, toluene, SDS and butanol, the smallest wavevector at which a

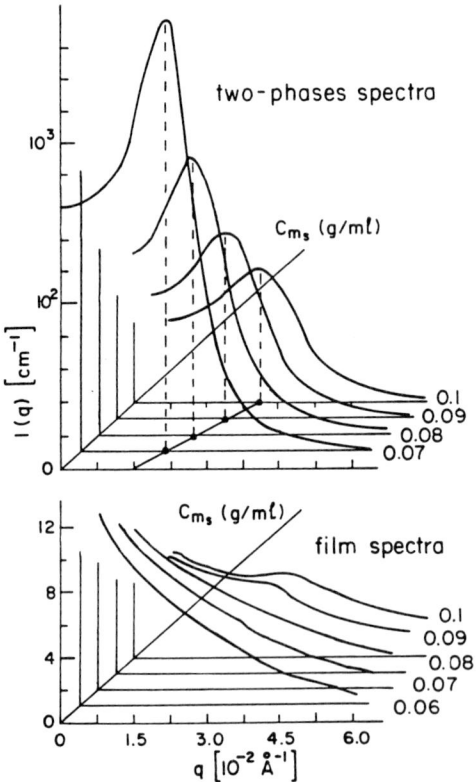

Fig. 1.11 Neutron scattering intensities from the system brine, toluene and sodium dodecyl sulfate (SDS) as given by Auvray *et al.* (1984). The volume fraction of brine and oil are equal. The mass density of amphiphile is denoted C_{ms}. The upper spectra are proportional to the water–water structure function, the lower to the amphiphile–amphiphile structure function. Reprinted with permission from L. Auvray, J.-P. Cotton, R. Ober, and C. Taupin, *J. Phys. Chem.* **88**, 4586. Copyright 1984 American Chemical Society.

peak is shown is about 1.5×10^{-2}Å indicating structure with a scale of about 400 Å. As the concentration of the amphiphile (denoted C_{ms} in the figure) is increased, the wavevector at the maximum increases, and the magnitude of the peak decreases. At large wavevectors, the water–water structure function falls off as q^{-4}, the signal expected from a system with extensive internal interface (Porod, 1951; Debye et al., 1957). A simple form incorporating this limit was proposed by Teubner and Strey (1987)

$$S(\mathbf{q}) = (a + gq^2 + cq^4)^{-1}, \tag{1.2}$$

which, with a negative value for g, fits the data from scattering experiments very well. By appropriately deuterating the sample, neutron scattering can also determine the amphiphile–amphiphile structure function, which is shown in the lower part of Fig. 1.11. Much less structure, if any, is revealed here.

One expects the structure revealed in the water–water structure function to consist of micelles when examining the phase with most amphiphile in the two-phase region, or the middle phase near either critical endpoint. The reason is that, in both cases, there is a much larger fraction of oil than of water, or *vice versa*. Micelles only form when there is a sufficient concentration of amphiphile, of course, and they do so when the loss of entropy of mixing of a molecular dispersion is balanced by the gain of energy on forming the micelle. The concentration of amphiphile at which micelles form is referred to as the critical micelle concentration, or cmc (Lang and Morgan, 1980; Mitchell et al., 1983). It is observed in experiment (Kahlweit et al., 1990) that at the critical endpoints, the critical fluid is a micellar solution. In the spectator phase with which it is in equilibrium, the amphiphiles are molecularly disperse.

We now imagine increasing the minority component, either oil or water, until it spans the system as does the majority component. The middle phase is now bicontinuous, and we expect that the amphiphile will no longer make micelles, but rather sheets which span the system, separating water from oil. The continuity of the water fraction can be examined by measuring the conductivity of the sample (Lagues and Sauterey, 1980; Porte et al., 1988b; Peyrelasse and Boned, 1990) which shows a rapid rise at a point at which, presumably, the water spans the system. The bicontinuous structure can also be demonstrated by measuring the diffusivities of oil, water and amphiphile (Guéring and Lindman, 1985; Olsson et al., 1986; Bodet et al., 1988a, b; Kahlweit et al., 1987). A typical result is shown in Fig. 1.12. At low salt concentrations one has a nearly pure oil phase in equilibrium with a water-rich phase which contains most of the amphiphile. This phase contains micelles with oil inside of them. One sees that the diffusion coefficient of water is very large as it spans the system. That of the oil and amphiphile, SDS, are about the same as they are locked within the same structure, and both are

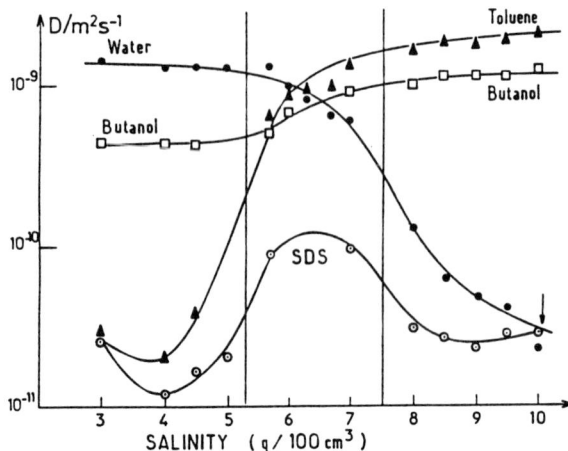

Fig. 1.12 Self-diffusion coefficients of oil-, water- and amphiphile-molecules, as a function of the salinity, in the system water, salt, toluene, SDS and butanol. Reprinted with permission from P. Guéring and B. Lindman, *Langmuir* **1**, 464. Copyright 1985 American Chemical Society.

much smaller because the micelles diffuse much more slowly than molecular water. As the concentration of salt is increased, one sees that the diffusion coefficients of the oil rises to a value similar to that of the water; both now span the system so the middle phase is bicontinuous. The diffusion coefficient of the amphiphile also increases. Finally, at large salt concentrations in which the system returns to two phases, a nearly pure water phase and an oil phase in which most of the amphiphile appears in the form of inverted micelles with water within, one sees that, in the latter, the diffusion coefficient of oil remains large, but that of the water and amphiphile are about the same and much smaller.

The most direct visualization of the structure of the middle phase comes from freeze-fracture microscopy (Bodet *et al.*, 1988a; Jahn and Strey, 1988; Vinson *et al.*, 1991). A picture taken of the strong amphiphile system $C_{12}E_5$, water, octane at a ratio of oil/(oil + water) of 0.4 is shown in Fig. 1.13. The stippled regions are oil, the smooth ones are water. The bar in the figure shows a distance of 2000 Å. One sees the incredibly convoluted pattern of oil and water in this fluid which certainly merits it the adjective "complex"!

1.3   Theoretical approaches to amphiphilic systems

In the formulation of any theory of physical phenomena, one must decide at the outset what the basic physical quantities of the theory are to be; that is, the

Fig. 1.13 Structure of the microemulsion, as obtained from freeze-fracture microscopy. Bar is 2000 Å. From Jahn and Strey (1988).

fundamental length and energy scales must be established. Different theories of amphiphilic systems have emerged which are distinguished by three different choices of the basic length and energy scales. We outline them briefly below.

1.3.1 Microscopic models

The distinguishing feature of microscopic models is that the basic statistical entities are taken to be the molecular components of the system, so that the fundamental scales are molecular ones. The mesoscopic lengths observed in scattering experiments and micrographs must then be derived from the theory. As all such theories to date have attempted to describe generic behaviour, the distinguishing features of particular amphiphiles and oils are ignored. All components are replaced by statistical objects with little or no internal structure whose interactions can be reduced to a few parameters or functions. These interactions are invariably of short range. They are attractive between polar entities, those representing the water and the head groups of the amphiphile, and between non-polar entities, those representing the oil and the tails of the amphiphile. They are repulsive between polar and non-polar objects, an interaction which reflects the fact that a polar and a non-polar entity can not lower their energy in one another's presence to the extent that two polar objects can. The amphiphile has been treated as a scalar particle without internal structure, or more commonly, as a vector whose interaction energies with the other components depends on whether it is aligned so that the polar and non-polar entities are close to one another or not. In a few cases, the amphiphile has been given more structure, and treated as a short copolymer consisting of a string of polar groups in the head and non-polar groups in the tail. The water molecule is most often treated as an object without internal structure, but occasionally it is represented by a vector. This is done so that interactions between water molecules and water and amphi-philic molecules can mimic the directional hydrogen-bonding forces, which are thought to be responsible for the lower critical point observed in the binary water–amphiphile system. The oils are treated as structureless.

An additional simplification that is often made is to replace the continuum system by a discrete one, which then reduces the description to a form of lattice fluid.[†] There are several advantages to this approach. Bulk phase behaviour and interfacial behaviour of lattice fluids have been studied for many years, so these properties of the amphiphilic system are well within the purview of such theories. Further, as the basic scale is at the molecular level, one can study the progression from weak to strong amphiphilic systems, and observe the corresponding changes in the bulk phase behaviour, the structure of the interfaces, their tensions, and the associated wetting properties. The self-assembly of the constituents into various structures, such as micelles and sheets can also be followed. Ordered arrays of these structures, as occurs in lyotropic phases, can be studied as straightforwardly as are ordered struc-tures in other systems; the disordered array in the microemulsion can be probed by the calculation of structure factors, and of the continuity of the various components. The real space structure of micelles, sheets and the microemulsion itself, can be examined by Monte Carlo simulation.

1.3.2 Ginzburg–Landau theories

Theories of this kind describe the system in terms of order parameters which represent microscopic quantities averaged over small regions, perhaps several molecular diameters. In principle, a Ginzburg–Landau free energy, which is an expansion in powers of the order parameters, can be derived from a microscopic theory, but more often it is simply constructed from symmetry arguments, once the choice of order parameters to be employed to describe the system has been made. As a result, it produces generic behaviour which depends upon the unknown coefficients in the expansion. The advantage of such theories is that they are simple enough to permit largely analytic analysis. Bulk and interfacial behaviour are easily described in this approach, as are structures such as spherical and cylindrical micelles. Phases consisting of ordered arrays of these objects or of amphiphilic sheets can also be encom-passed, much as they are in similar theories of type II superconductivity.

[†]For a review of lattice models see Gompper and Schick (1994a).

Structure functions of disordered phases are easily calculated. Again, the mesoscopic length of the microemulsion must emerge from the theory, as it is much larger than the input length. Monte Carlo simulations provide a real-space description whose scale is, again, somewhat larger than that of microscopic theories.

1.3.3 Membrane theories

Strong amphiphiles are relatively insoluble in oil and water, and form monolayers at oil/water interfaces in ternary mixtures. These monolayers can be considered to act as nearly incompressible two-dimensional liquids with a characteristic area per head group determined by the molecular interaction between amphiphiles. In binary systems, the amphiphiles also form sheets of constant area, but they now exist in the form of bilayers. Membrane theories take as the basic statistical entities these sheets of amphiphile, whether monolayer or bilayer, thereby providing a universal description for ternary and binary mixtures. The length and energy scales are not the molecular size of the head group or the molecular energy involved in assembling the amphiphilic sheet. Rather the basic energy is that needed to bend the sheet, an energy which arises microscopically from a change in the local concentration of head and tail groups of the amphiphiles. Over short distances, this energy will ensure that the membrane is relatively flat. However, over larger distances, the effect of thermal fluctuations is to cause the orientation of the membrane to change appreciably, so that a knowledge of the orientation at one position imparts no knowledge as to its orientation a large distance away. The scale of distance at which this crossover occurs is called the persistence length, and it can be calculated from the free energy describing a system of sheets. The local free energy per unit area is written as an expansion in powers of the local curvatures, and the integrated free energy depends parametrically on the unknown coefficients in this expansion. These coefficients are the spontaneous, or natural, curvature of the membrane, and its bending and saddle-splay moduli. The persistence length is found to depend exponentially on the bending modulus, and therefore can vary over several orders of magnitude. For very stiff membranes, this basic length can be of the order of microns, whereas for more pliant ones it is appreciably less. In all cases, the persistence length is orders of magnitude larger than molecular lengths, and it is identified with the mesoscopic length characteristic of the disordered microemulsion.

The membrane approach is most useful for the study of the fluctuations of a single membrane or those characterizing a bulk phase, and the scaling behaviour of such phases is readily obtained. One can also determine straightforwardly the relative stability of different ordered conformations of sheets, such as are found in the lyotropic phases, or which might underlie the structure of the microemulsion. The most difficult phase to study in this approach is the disordered microemulsion as it requires a statistical mechanics of random surfaces. Monte Carlo simulation of more than one sheet is quite difficult, and there has been no such application as yet to the disordered microemulsion.

2 Microscopic models

2.1 Formulation

2.1.1 The three-component model

As noted earlier, the phase behaviour of weak amphiphilic systems is that expected from an ordinary ternary mixture. Thus the Hamiltonian of such a mixture is an appealing zeroth-order approximation for the amphiphilic system. In a lattice model in which all components are found on the sites, this Hamiltonian is simply

$$\mathcal{H}_0 = -\sum_{\alpha\beta}\sum_{\langle ij\rangle} E^{\alpha\beta} P_i^\alpha P_j^\beta - \sum_\alpha\sum_i \mu^\alpha P_i^\alpha, \qquad (2.1)$$

where $P_i^\alpha = 1$ if there is a molecule of species α on site i, and is zero otherwise, and α takes on the values a, b and c representing water, oil and amphiphile respectively. There is a molecule at every site,

$$\sum_\alpha P_i^\alpha = 1. \qquad (2.2)$$

For simplicity, we have restricted ourselves to interactions only between nearest neighbours; the notation $\langle ij\rangle$ indicates that the sum is over all such distinct pairs. This simple Hamiltonian has been studied extensively (Sivardiere and Lajzerowicz, 1975; Furman et al., 1977). It is readily mapped to a spin-one model of magnetism via the relations

$$P_i^a = (1 + S_i)S_i/2,$$

$$P_i^b = -(1 - S_i)S_i/2, \qquad (2.3)$$

$$P_i^c = (1 - S_i^2),$$

where the S_i take on the values $1, 0, -1$. Upon substitution of these relations into (2.1) we obtain, up to a constant, the spin Hamiltonian of the Blume–

Emery–Griffiths (BEG) model (Blume *et al.*, 1971)

$$\mathcal{H}_{BEG} = -\sum_{\langle ij \rangle}[JS_iS_j + KS_i^2S_j^2 + C(S_i^2S_j + S_iS_j^2)] - \sum_i (HS_i - \Delta S_i^2),$$

$$(2.4)$$

where

$$4J = E^{aa} + E^{bb} - 2E^{ab},$$

$$J + K + 2C = E^{aa} + E^{cc} - 2E^{ac},$$

$$J + K - 2C = E^{bb} + E^{cc} - 2E^{bc},$$

$$2H = \mu^a - \mu^b - 2d(E^{ac} - E^{bc}),$$

$$2\Delta = 2\mu^c - \mu^a - \mu^b + 2d(E^{ac} + E^{bc} - 2E^{cc}), \qquad (2.5)$$

with d the dimensionality of the cubic lattice. The right-hand sides of the first three equations above are proportional to the critical temperatures of the three binary mixtures. Because the critical temperature of the oil–water system is certainly much higher than than of the other two, one should choose $3J \gg K, |C|$.

There is little in this Hamiltonian to differentiate the amphiphile from the other two components. Its tendency to form interfaces between oil and water can be incorporated into the model by extending the Hamiltonian of (2.1) to (Schick and Shih, 1987)

$$\mathcal{H} = -\sum_{\alpha\beta}\sum_{\langle ij \rangle} E^{\alpha\beta} P_i^\alpha P_j^\beta - \sum_\alpha \sum_i \mu^\alpha P_i^\alpha$$

$$- L\sum_{(ijk)} (P_i^a P_j^c P_k^a + P_i^b P_j^c P_k^b - P_i^a P_j^c P_k^b - P_i^b P_j^c P_k^a), \qquad (2.6)$$

where the sum in the third term is over all sets of three sites in a row. With $L < 0$, the new term reduces the energy of a configuration in which an amphiphile sits between a water and oil by an amount $|L|$, and increases by the same amount the energy of configurations in which the amphiphile sits between two water or two oil molecules. In magnetic language, the Hamiltonian now reads

$$\mathcal{H} = -\sum_{\langle ij \rangle}[JS_iS_j + KS_i^2S_j^2 + C(S_i^2S_j + S_iS_j^2)]$$

$$- \sum_i (HS_i - \Delta S_i^2) - L\sum_{(ijk)} S_i(1 - S_j^2)S_k. \qquad (2.7)$$

It is not difficult to see that the additional interaction, which is antiferromagnetic between sites i and k, favours lyotropic phases such as the lamellar configuration $+0-0+0-0$. As this has a period of four, the wavevector which describes it does not touch the Brillouin zone edge, so that it is not singled out by a symmetry of the underlying imposed lattice. Hence we expect to find many different lamellar states characterized by different wavevectors, many of long period, just as occur in other lattice models with competing interactions (Bak, 1982; Fisher and Huse, 1982; Selke, 1988, 1992).

One can view the addition of the above three-particle interaction as a way of including the most obvious property of amphiphiles, their tendency to sit between oil and water, without having to introduce additional degrees of freedom associated with their directional properties. Alternatively, such multi-particle interactions can be looked upon as resulting from integrating out these directional properties, a point of view we shall make explicit in Section 2.1.4 below.

The three-particle interaction of (2.6) does not describe the tendency to form amphiphilic bilayers in the binary system with water. To do this, an analogous interaction between four particles can be introduced (Gompper and Schick, 1989b),

$$\mathcal{H}_4 = -W \sum_{i,\delta} P^a_{i-\delta} P^c_i P^c_{i+\delta} P^a_{i+2\delta}, \tag{2.8}$$

which mimics the tail–tail and the head–head attraction, and therefore favours the formation of bilayers of amphiphile in water-rich or oil-rich phases. The competition between bilayer and monolayer formation as a function of the oil/water ratio could be studied if both the three- and four-particle interactions were incorporated into the model. Most studies, however, have focused on the Hamiltonian of (2.6) suited to the ternary system.

2.1.2 The Widom model

This model, a generalization of one introduced earlier by Wheeler and Widom (1968), shares with the three-component model the feature that the directional properties of the amphiphile ultimately do not appear as statistical variables in the Hamiltonian. This makes it relatively easy to study. In order to place this model in context, however, we will begin with a form in which the directional properties are explicit, and then will eliminate them (Hofsäss and Kleinert, 1988).

There are again three kinds of molecules, a, b and c, representing water, oil and amphiphile. All molecules occupy the bonds of a cubic lattice. The water and oil are treated as structureless. The interaction between oil and water is taken to be infinitely repulsive, so that they can never be in contact. The part

of the Hamiltonian which describes only the oil and water and their interactions is simply

$$\mathcal{H}_1 = - \sum_{\langle ij \rangle \langle jk \rangle} (E^{aa} P_{ij}^a P_{jk}^a + E^{bb} P_{ij}^b P_{jk}^b) - \sum_{\langle ij \rangle} (\mu^a P_{ij}^a + \mu^b P_{ij}^b), \qquad (2.9)$$

where $P_{ij}^\alpha = 1$ if there is a molecule of species α on the bond $\langle ij \rangle$, and vanishes
otherwise. The amphiphile on bond $\langle ij \rangle$ can take two directions: parallel or
antiparallel to the vector from site i to j. This can be specified by introducing
the operators Q_{ij}^+ which are equal to unity if the head of the amphiphile on
bond $\langle ij \rangle$ points to site i, and are zero otherwise, and Q_{ij}^- which is equal to
unity if the tail points to site i, and vanishes otherwise. The interaction
between water and the head group of the amphiphile is $-E^{aa'}$, and that
between the oil and tail group is $-E^{bb'}$. The interactions between the water
and tail group and oil and head group are infinitely repulsive. Thus the part of
the Hamiltonian describing the interactions of amphiphile with oil or water is

$$\mathcal{H}_2 = - \sum_{\langle ij \rangle \langle jk \rangle} [E^{aa'} (P_{ij}^a P_{jk}^c Q_{jk}^+ + P_{ij}^c Q_{ij}^- P_{jk}^a)$$
$$+ E^{bb'} (P_{ij}^b P_{jk}^c Q_{jk}^- + P_{ij}^c Q_{ij}^+ P_{jk}^b)]. \qquad (2.10)$$

Finally, the interactions between amphiphile head groups is $-E^{a'a'}$, while that
between tail groups is $-E^{b'b'}$. The interaction between head and tail groups is
infinitely repulsive. The Hamiltonian describing the amphiphiles is, therefore,

$$\mathcal{H}_3 = - \sum_{\langle ij \rangle \langle jk \rangle} [E^{a'a'} P_{ij}^c Q_{ij}^- P_{jk}^c Q_{jk}^+ + E^{b'b'} P_{ij}^c Q_{ij}^+ P_{jk}^c Q_{jk}^-]$$
$$- \mu^c \sum_{\langle ij \rangle} P_{ij}^c. \qquad (2.11)$$

The total Hamiltonian is

$$\mathcal{H} = \mathcal{H}_1 + \mathcal{H}_2 + \mathcal{H}_3. \qquad (2.12)$$

Note that, by virtue of the infinite repulsions, all water and oil molecules *must*
be separated by amphiphiles, and all amphiphiles *must* be situated properly,
with head groups pointing to other head groups or to water, tails to other tails
or to oil. It is understood that only configurations obeying these constraints
are to be included in the partition function constructed from the Hamiltonian
of (2.12).

The simplicity of the Widom model lies in the fact that it can be mapped
directly to a spin-$\frac{1}{2}$ Ising model in which only allowed configurations can
possibly appear; that is, configurations of Ising spins are in a one-to-one
correspondence with allowed configurations of the original molecules. The

mapping is

$$P_{ij}^a = (1 + \sigma_i)(1 + \sigma_j)/4,$$

$$P_{ij}^b = (1 - \sigma_i)(1 - \sigma_j)/4,$$

$$P_{ij}^c Q_{ij}^+ = (1 + \sigma_i)(1 - \sigma_j)/4,$$

$$P_{ij}^c Q_{ij}^- = (1 - \sigma_i)(1 + \sigma_j)/4, \tag{2.13}$$

where $\sigma_i = \pm 1$. In this representation, each molecule is represented by two spins, one at each end of the bond. Two up spins represent a water, two down spins an oil, and an up–down pair represents an amphiphile with the up spin as the head group. A single spin on a given site is part of all molecules on the bonds leading to that site. One readily verifies in the spin representation that the four operators on the left are indeed projection operators, as they were defined to be, and that their sum is unity so that there is one particle of some kind on each bond. If these identities are substituted into the total Hamiltonian, one obtains that of an equivalent Ising model (Widom 1986)

$$\mathcal{H}_W = - h \sum_i \sigma_i - J \sum_{\langle ij \rangle} \sigma_i \sigma_j - M \sum_{\langle ik \rangle}' \sigma_i \sigma_k$$

$$- 2M \sum_{\langle ik \rangle}'' \sigma_i \sigma_k - L \sum_{\langle ijk \rangle} \sigma_i \sigma_j \sigma_k, \tag{2.14}$$

where the third sum is over all distinct pairs of spins a distance of two lattice constants apart, the fourth sum is over all distinct pairs a distance $\sqrt{2}$ lattice spacings apart, and the final sum over all distinct contiguous triplets of spins. The even interactions are

$$J = \frac{(z - 1)}{4}(E^{aa} + E^{bb} - 2E^{cc}) + \frac{(\mu^a + \mu^b - 2\mu^c)}{4}, \tag{2.15}$$

$$M = \tfrac{1}{8}[(E^{aa} + E^{a'a'} - 2E^{aa'}) + (E^{bb} + E^{b'b'} - 2E^{bb'})], \tag{2.16}$$

while the symmetry-breaking fields are

$$L = \tfrac{1}{8}[(E^{aa} + E^{a'a'} - 2E^{aa'}) - (E^{bb} + E^{b'b'} - 2E^{bb'})], \tag{2.17}$$

$$h = \frac{z(z - 1)}{16}[3(E^{aa} - E^{bb}) + 2(E^{aa'} - E^{bb'})$$

$$- (E^{a'a'} - E^{b'b'})] + \frac{z}{4}(\mu^a - \mu^b), \tag{2.18}$$

with z the coordination number of the lattice.

A comparison with the parameters of the three-component model is instructive. From the three parameters J, M and L of the Widom model and from J, K and C of the three-component model one obtains the strengths of the interactions within the three binary systems. The scale of the oil–water phase separation is clearly set by the parameters J in each model. Similarly, the difference in the two transition temperatures of the amphiphile–water system and amphiphile–oil system are given by L of the Widom model and C of the three-component model respectively. The sum of these two transition temperatures is given by M in the former and $J + K$ in the latter model. The signs of $J + K$ and of M are important. The above interpretation in terms of transition temperatures clearly indicates that they are positive. If M is positive (ferromagnetic in magnetic language) then it is easy to see that it will tend to produce lamellar phases of the form $+ - + -$. Such period-two phases are a direct consequence of the use of a lattice, unfortunately, a result which is seen from the fact that the wavevector which characterizes these phases touches the Brillouin zone boundary at a point of high symmetry. (In technical terms, the Lifshitz condition is fulfilled; see Lifshitz and Pitaevskii (1980). This does not occur in the three-component model because the same phase there has a period four.) As a consequence, one expects no lamellar phases of long wavelength to appear in the Widom model with M positive.

The derivation by Widom (1986) of the above model is different and corresponds to the choice $E^{aa} = E^{bb} = E^{aa'} = E^{bb'} = 0$; that is, the only non-vanishing interactions besides the infinite repulsions are those between amphiphiles. Further, the interaction between amphiphiles is taken to be repulsive, $E^{a'a'} + E^{b'b'} < 0$ so that, from (2.16) above, $M < 0$. Other interpretations of a negative M are possible, however. From (2.16), one sees that this sign would also arise if the attractive interaction between water and the polar amphiphile head were stronger than the average of the interactions between waters and between polar head groups, and similarly for the interaction between the tails and oil. In any case, a negative value for M favours lamellar phases such as $+ + - - + + - -$ in which the amphiphiles are further apart. As these are period-four phases, their existence is not tied to a symmetry of the underlying lattice and one may expect to find long-period phases with wavevectors incommensurate with that of the lattice. Such phases have been well studied in the anisotropic axial next-nearest-neighbour Ising (ANNNI) model (Bak, 1982; Fisher and Huse, 1982; Selke, 1988, 1992), of which the Widom model with $M < 0$ is a variant. It is with this sign of M that the Widom model has been much studied.

2.1.3 The Alexander model

In the three-component model, all molecules are placed on lattice sites; in the Widom model, they are all placed on bonds. In the Alexander model, however, the molecules are not treated symmetrically. Oil and water are placed on sites, and every site is occupied by one of the two molecules; amphiphiles are placed on bonds, but bonds can also be empty (Alexander, 1978). The Hamiltonian of the system is taken to be (K. Chen *et al.*, 1987, 1988; Stockfisch and Wheeler, 1988)

$$\mathcal{H} = -\sum_{\langle ij \rangle} \sum_{\alpha\beta} E^{\alpha\beta} P_i^\alpha P_j^\beta - \sum_i \sum_\alpha \mu^\alpha P_i^\alpha$$

$$-\sum_{\langle ij \rangle \langle jk \rangle} E^{cc} P_{ij}^c P_{jk}^c - \sum_{\langle ij \rangle \langle ik \rangle}' \tilde{E}^{cc} P_{ij}^c P_{jk}^c - \sum_{\langle ij \rangle} \mu^c P_{ij}^c$$

$$-\sum_{\langle ij \rangle} \sum_{\alpha\beta} E^{\alpha c \beta} P_i^\alpha P_{ij}^c P_j^\beta, \tag{2.19}$$

where α and β take the values a and b, and

$$\sum_\alpha P_i^\alpha = 1. \tag{2.20}$$

The first line of the Hamiltonian contains the nearest-neighbour interactions within the binary oil and water system and the associated chemical potentials. The second line describes the interactions between pairs of amphiphiles on adjacent bonds which make an angle of π or $\pi/2$ with one another, and also contains the amphiphile chemical potential. The coupling between these two systems is provided by the final term, a three-particle interaction which occurs when an amphiphile sits between two other molecules. One sees from the constraint of (2.20) that, in the two-component limit, the water (or oil) becomes completely trivial and decouples from the amphiphiles. Thus the model is restricted to a description of ternary mixtures with comparable amounts of oil and water. It is also clear that attractive interactions between amphiphiles will tend to produce lamellar phases of period two, just as in the Widom model. Thus there will be no long-period lamellar phases. However, if the interactions are repulsive, such phases are to be expected.

2.1.4 Vector models

To this point, we have considered microscopic models in which the amphiphile has essentially no orientational degrees of freedom. Several lattice models which include these degrees of freedom have been studied. Almost all of them begin with the Hamiltonian of a simple ternary mixture, and with particles on the sites of a lattice. The Hamiltonian is then $\mathcal{H}_0$ of (2.1) which we repeat here:

$$\mathcal{H}_0 = -\sum_{\alpha\beta}\sum_{\langle ij\rangle} E^{\alpha\beta} P_i^\alpha P_j^\beta - \sum_\alpha\sum_i \mu^\alpha P_i^\alpha. \tag{2.21}$$

To this is added a Hamiltonian, $\mathcal{H}_{amp}$, which accounts for the orientation-dependent interactions. It is the different choice for this Hamiltonian which accounts for the several different vector models of amphiphilic systems.

One of the earlier models of this kind was proposed for a binary water–amphiphile system by Halley and Kolan (1988). Generalized to the ternary system by Gompper and Schick (1989b), the additional part of the Hamiltonian reads

$$\mathcal{H}_{amp} = \sum_{\langle ij\rangle}(\gamma_2 P_i^a P_j^c - \gamma_1 P_i^b P_j^c)\boldsymbol{\tau}_j \cdot \mathbf{r}_{ij}, \tag{2.22}$$

where $\mathbf{r}_{ij}$ is a unit vector in the direction from site i to site j, and the orientation of an amphiphile on site i is characterized by the unit vector $\boldsymbol{\tau}_i$. With positive γ_1 and γ_2, this favours configurations in which the amphiphile is properly oriented with respect to oil and water. If there are no other orientational interactions, the angular degrees of freedom can be integrated out to produce an effective, temperature-dependent interaction,

$$\mathcal{H}_{amp} = \sum_i \ln[\sinh(\beta V_i)/\beta V_i], \tag{2.23}$$

where $\beta = 1/k_B T$, and

$$V_i = \frac{1}{2} P_i^c \left\{ \sum_{\delta=1}^3 \left[(\gamma_1 + \gamma_2)(P_{i+\delta}^a - P_{i+\delta}^b - P_{i-\delta}^a + P_{i-\delta}^b) \right. \right.$$
$$\left. \left. -(\gamma_2 - \gamma_1)(P_{i+\delta}^c - P_{i-\delta}^c) \right]^2 \right\}^{1/2}, \tag{2.24}$$

where δ is a vector between nearest-neighbour sites. This interaction can be expanded in a power series in the projection operators P_i, a series which terminates because $P_i^2 = P_i$. The resulting form is that of a sum of multi-particle interactions between a given amphiphile *without* orientational degrees of freedom and all combinations of its nearest neighbours. One of these temperature-dependent interactions is of the form of the three-particle interaction of (2.6). In this way, the three-component model can be viewed as incorporating some of the effects of the orientation-dependent interactions of the amphiphile without having to introduce additional degrees of freedom to describe them.

In contrast to the continuous orientational degree of freedom above, most other models permit the vector amphiphile to point only to nearest-neighbour sites. A rather general form which accounts for interactions between a simple vector amphiphile and structureless water and oil molecules is given by Matsen and Sullivan (1990). The directions along which the vector can point are restricted to those of an underlying cubic lattice. Matsen and Sullivan further consider nearest-neighbour interactions only and systems which are symmetric under the interchange of oil and water and the two ends of the amphiphile (i.e. to a balanced system). Under these restrictions, the most general form of the Hamiltonian can be written as

$$\mathcal{H} = \mathcal{H}_0 + \mathcal{H}_{amp}, \tag{2.25}$$

$$\mathcal{H}_{amp} = -\sum_{\langle ij \rangle} \sum_{\nu=2}^{4} [J_\nu Q_{\nu,ij} + K_\nu Q_{\nu,ij}^2], \tag{2.26}$$

where the interaction between the amphiphile and oil and water is governed by

$$Q_{2,ij} = (P_i^a - P_i^b) P_j^c (\tau_j \cdot \mathbf{r}_{ji}) + P_i^c (P_j^a - P_j^b)(\tau_i \cdot \mathbf{r}_{ij}), \tag{2.27}$$

that between amphiphiles parallel or antiparallel to one another and to the vector joining them is given by

$$Q_{3,ij} = P_i^c P_j^c (\tau_i \cdot \mathbf{r}_{ij})(\tau_j \cdot \mathbf{r}_{ji}), \tag{2.28}$$

and that between amphiphiles perpendicular to the vector joining them by

$$Q_{4,ij} = -P_i^c P_j^c (\tau_i \times \mathbf{r}_{ij}) \cdot (\tau_j \times \mathbf{r}_{ji}). \tag{2.29}$$

In each case, the parameter K_ν gives the average energy of the interaction and $2J_\nu$ the difference between most favourable and unfavourable orientations. The binary amphiphile–water limit can also be treated within this model by restricting $P_i^b = 0$. If the interactions between amphiphiles themselves, J_3, K_3, J_4, K_4, are set to zero, this model reduces to one considered by Ciach et al. (1988, 1989, 1991) and Laradji et al. (1991a). In this case, the vector degrees of freedom could again be summed out leading to multi-particle interactions with temperature-dependent couplings. This has actually only been carried out in one and two dimensions (Skaf and Stell, 1992a; Slotte, 1992).

An extension of the above which treats the vector amphiphiles as continuous rather than discrete variables, and which retains interactions between the amphiphiles, has been considered by Gunn and Dawson (1992). The most general Hamiltonian for such a system requires an infinite number of interaction parameters, that is, interaction functions of the continuous angle $\tau \cdot \mathbf{r}$. To avoid this difficulty, Gunn and Dawson (1992) assume that the amphiphile always points in a direction close to those of nearest neighbours, and expand the interactions to second order in the quantities $\tau \cdot \mathbf{r}$, small for τ nearly perpendicular to $\mathbf{r}$, and $(1 - \tau \cdot \mathbf{r})$, small for τ nearly parallel to $\mathbf{r}$. Thus, their model can be considered to be a perturbation about the above discrete model. For their actual calculation, they further assume that in interactions with its own kind, the amphiphile is structureless, i.e. such interactions do not distinguish between heads and tails. This assumption may be adequate near the balanced system, the one considered by them for the most part. It would fail in the two-component limit where the formation of bilayers is important.

All the above models explicitly consider the orientational interactions between two amphiphiles, and between an amphiphile and a structureless water or oil molecule. The water molecule has a dipole moment, of course, one important in hydrogen bonding. By treating the water as structureless, these models miss orientational bonding effects between water molecules and between water and amphiphile molecules. The latter are thought to be responsible for the lower critical point displayed by the binary water–amphiphile system (Walker and Vause, 1983; Goldstein, 1985). Without such a lower critical point, the phase behaviour of the non-ionic amphiphilic systems cannot be reproduced. With an eye to obtaining a realistic phase diagram for these systems, Matsen $et\ al.$ (1993) have assigned vector internal degrees of freedom, τ, to $both$ the amphiphilic and water molecule. These degrees of freedom are discrete, pointing only to nearest neighbours. The Hamiltonian is

$$\mathcal{H} = \mathcal{H}_0 + \mathcal{H}_{amp}, \tag{2.30}$$

with

$$
\begin{aligned}
\mathcal{H}_{amp} = -\sum_{\langle ij \rangle} [& P_i^a P_j^c (L_a \delta(\mathbf{r}_{ji}, \tau_j) + L_c \delta(\mathbf{r}_{ij}, \tau_i) \\
& + L_{ac} \delta(\mathbf{r}_{ji}, \tau_j) \delta(\mathbf{r}_{ij}, \tau_i) - E^{ac}) \\
& + P_j^a P_i^c (L_a \delta(\mathbf{r}_{ij}, \tau_i) + L_c \delta(\mathbf{r}_{ji}, \tau_j) \\
& + L_{ac} \delta(\mathbf{r}_{ij}, \tau_i) \delta(\mathbf{r}_{ji}, \tau_j) - E^{ac}) \\
& + P_i^b P_j^c (L_b \delta(\mathbf{r}_{ij}, \tau_j) - E^{bc}) \\
& + P_j^b P_i^c (L_b \delta(\mathbf{r}_{ji}, \tau_i) - E^{bc})],
\end{aligned}
\tag{2.31}
$$

with $\delta(a, b)$ the Kronecker delta. The coefficients L_a, L_b, etc. give the strengths of the interactions between various orientations of the molecules. There are three major differences between this Hamiltonian and that of (2.26). First, this Hamiltonian does not treat the oil and water symmetrically, so that the interaction between the amphiphile and water can be different from that between amphiphile and oil. This is certainly necessary in order to produce a phase diagram similar to that of experimental systems in which the upper critical points of the two binary systems containing amphiphile are at very different temperatures. Second, this Hamiltonian contains a term, with coefficient L_{ac}, which gives the additional strength of the amphiphile–water interaction if the two point to one another. Because a large number of possible orientations is desirable in the description of hydrogen bonding (Walker and Vause, 1983), an fcc lattice with twelve nearest-neighbour interactions is chosen. Third, while this Hamiltonian includes the orientational interactions between amphiphile and water which are neglected in that of (2.26), it does not include the orientational interactions between amphiphiles themselves which are described there. Clearly it would be desirable to study a Hamiltonian with both kinds of orientational interactions.

2.1.5   Other microscopic models

All of the above models are defined on a lattice, and treat the amphiphile as having few, if any, degrees of freedom. There have been a few attempts to treat the system as a continuous one, and to assign more complexity to the amphiphile. The first of these considers the effect of the simplest interactions between amphiphile and water or oil in a continuum model of a ternary mixture (Telo da Gama and Gubbins, 1986; Telo da Gama and Thurtell, 1986; Telo da Gama, 1987). All molecules are taken to be hard spheres with isotropic van der Waals interactions between all components. In addition, there is a directional dependence of the interaction of amphiphile with both oil and water which varies with the cosine of the angle between the orientation of the amphiphile and the direction to the oil or water. There is no directionally-dependent interaction between two amphiphiles. Thus, this model can be considered to be a continuum generalization of the discrete one with $\mathcal{H}_{amp}$ given by (2.22) above.

A similar, but more complicated, continuum model, in which there is a directionally-dependent interaction, has been considered by Gunn and Dawson (1989). The amphiphiles have elliptical shape, while the water molecules remain spherical. Interactions between waters are of the Lennard-Jones form, while those involving either one or two amphiphiles are variants of it. The system was studied by molecular dynamics simulation.

An effort to provide the amphiphile with more degrees of freedom and to mimic the experimental ability to increase the head and tail groups in the series C_iE_j has been formulated by Larson et al. (1985), and simulated extensively by Larson (1988, 1989, 1992). In this lattice model, there are three components but only two kinds of units: water-like and oil-like ones. The amphiphile is similar to a small copolymer with a number of oil-like molecules strung together to form the tail, and another number of water-like molecules to form the head. The structure within the various phases can be studied as a function of, *inter alia*, the number of such head and tail groups. A very similar model defined in the continuum has also been studied by simulation (Smit *et al.*, 1990, 1991, 1993).

2.2.   Some results of the three-component model

2.2.1   Phase behaviour

The Hamiltonian of (2.7) with the amphiphilic strength L set to zero is the Blume–Emery–Griffiths model (Blume *et al.*, 1971) which has been extensively studied, particularly within mean-field theory (Mukamel and Blume, 1974; Furman *et al.*, 1977). There are, in general, three phases which are respectively rich in oil, or water, or amphiphile. These three phases can coexist over a large region of the phase space. For our purposes, interesting phase behaviour is obtained at fixed temperature by varying the parameter C which, from (2.5), is related to the two-particle interactions by

$$4C = (E^{aa} + E^{cc} - 2E^{ac}) - (E^{bb} + E^{cc} - 2E^{bc}). \tag{2.32}$$

From this expression, we see that this parameter is related to the difference between the interaction of the amphiphile with water and with oil. When the amphiphile is ionic, it is thought that adding salt to the water affects just this difference. When $C = 0$, one finds within mean-field theory that there is a range of temperatures for which the system exhibits three-phase coexistence. If the temperature is fixed at a value at which this is so and the value of C is made to increase (amphiphile prefers oil), a positive value is attained at which three-phase coexistence ends at a critical end point. For larger values, only two phases coexist, one which is water-rich, the other a mixture of oil and amphiphile. Similarly, if C is made to decrease (amphiphile prefers water), a critical endpoint is reached beyond which there is only two-phase coexistence between an oil-rich phase and a mixture of water and amphiphile. Thus the sequence of two- to three- to two-phase coexistence as a parameter, like salt concentration, is varied is contained within this model. Further, this sequence remains when the amphiphilic interaction L is turned on (Gompper and Schick, 1990a). From this we conclude that the phase behaviour observed with changing salt concentration which is characteristic of oil, water, ionic-amphiphile systems has little or nothing to do with the strength of the amphiphile as measured by its ability to bring about micelles, lyotropic phases, etc.

   This same phase behaviour appears as a function of varying temperature in systems with a non-ionic amphiphile, and is intimately tied to the existence of a lower critical point in the binary water–amphiphile system (Kahlweit *et al.*,

1985). This lower critical point is thought to be due to hydrogen bonding between the amphiphile and water (Hirschfelder *et al.*, 1937) and is therefore a consequence of orientational degrees of freedom of these molecules. As discussed earlier in Section 2.1.4, such effects can only be encompassed in a model of molecules without orientation if the interaction between molecules is viewed as an effective, temperature-dependent one (Andersen and Wheeler, 1978; Walker and Vause, 1980; Goldstein, 1985; Hackenbroich, 1988). Carneiro and Schick (1988) demonstrated that by including this temperature dependence in the three-component model of the ternary system, the correct sequence of phases could be brought about as the temperature was varied. The behaviour was not very sensitive to the amphiphilic interaction L, from which we conclude once more that the phase behaviour has little to do with the strength of the amphiphile. This conclusion is in agreement with experiment (Kahlweit *et al.*, 1991). Additionally, the two critical end points can be made to vanish at a tricritical point as other interactions in the system are varied. Again this behaviour is in agreement with experiment (Kahlweit *et al.*, 1991).

We note at this point that the result that the pattern of phase behaviour is relatively insensitive to the strength of the amphiphile does not at all imply that the ability of the amphiphile to solubilize oil and water and to produce a fluid with structure is also independent of this strength. We will see below that this is not the case.

Let us now consider the phase diagram of the three-component model with temperature-independent interactions in the symmetric subspace, i.e. in which the amphiphile interacts equally strongly with water and oil which are present in equal amounts. This phase diagram has been calculated for a two-dimensional system (Gompper and Schick, 1989a, 1990b) via transfer-matrix[†] and Müller-Hartmann Zittartz methods (Müller-Hartmann and Zittartz, 1977). To remove a degeneracy in two dimensions, an additional term $-K_2 \sum S_i^2 S_j^2$, with the sum over distinct second neighbour pairs and K_2/J small, was added to the Hamiltonian of (2.7). The phase diagram is shown as a function of temperature, T, and amphiphile chemical potential, Δ, in Fig. 2.1. The phase diagram of the three-dimensional system is shown in Fig. 2.2. Both phase diagrams have been calculated for parameter values for which a lamellar phase is present.

For the three-dimensional system, the phase diagram has been calculated via mean-field theory, in which the density matrix $\hat{\rho}$ in the exact expression $F = \mathrm{Tr}[\hat{\rho}\mathcal{H}] + T\mathrm{Tr}[\hat{\rho}\ln\hat{\rho}]$ is replaced by a product of single-site density matrices. The Helmholtz free energy of the model (2.7) is given in this

---

[†]For a review see Barber (1984).

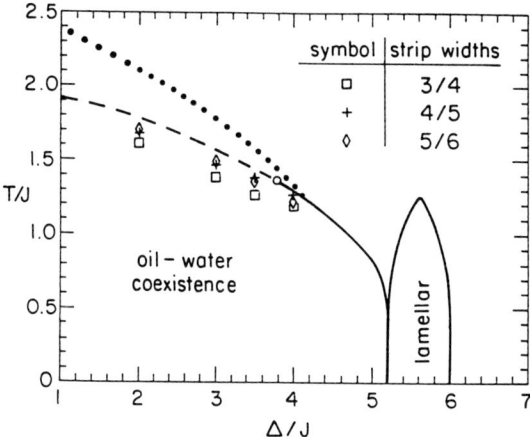

Fig. 2.1 Phase diagram of the three-component model in two dimensions as a function of temperature and amphiphile chemical potential, Δ. The concentrations of oil and water are equal. Interaction parameters are $K/J = 2$, $K_2/J = -0.2$, and $L/J = -3$. Full lines and dashed lines, results of a Müller-Hartmann Zittartz approximation, indicate first-order and continuous transitions respectively. The open circle denotes a tricritical point. Symbols show the results of transfer matrix calculations on strips of different widths. The dotted line is the disorder line (see Sect. 2.2.2 below). From Gompper and Schick (1989a).

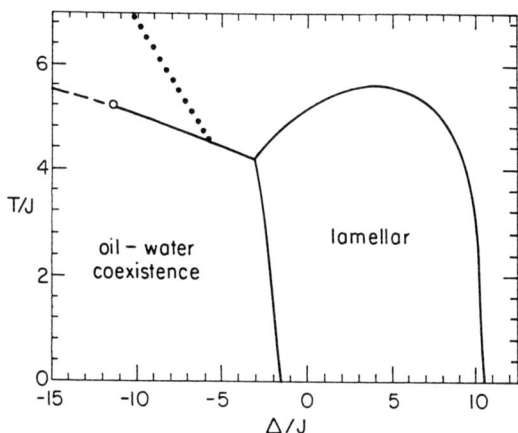

Fig. 2.2 Phase diagram of the three-component model in three dimensions as obtained from mean-field theory. The concentrations of oil and water are equal. Interaction parameters are $K/J = 0.5$, $L/J = -3.5$. Continuous and first-order transitions are denoted by dashed and solid lines respectively. The open circle denotes a tricritical point. The dotted line is the Lifshitz line (see Sect. 2.2.2 below). From Gompper and Schick (1990a).

approximation by (Gompper and Schick, 1990a)

$$\mathcal{F}_{MF}(\{M_i\}, \{Q_i\}, T) =$$

$$- \sum_{\langle ij \rangle} \left[JM_iM_j + KQ_iQ_j + C(M_iQ_j + M_jQ_i) \right]$$

$$- \sum_{(ijk)} LM_i(1 - Q_j)M_k \tag{2.33}$$

$$+ T \sum_i \left[\frac{Q_i + M_i}{2} \ln \frac{Q_i + M_i}{2} + \frac{Q_i - M_i}{2} \ln \frac{Q_i - M_i}{2} \right.$$

$$\left. + (1 - Q_i) \ln(1 - Q_i) \right]$$

with $M_i \equiv \langle S_i \rangle$ and $Q_i \equiv \langle S_i^2 \rangle$. These densities are obtained in terms of the fields H and Δ by constructing its Legendre transform, the Gibbs free energy,

$$\mathcal{G}_{MF}(H, \Delta, T) = \min_{M,Q} \left[\mathcal{F}_{MF}(\{M_i\}, \{Q_i\}, T) - \sum_i (HM_i - \Delta Q_i) \right], \tag{2.34}$$

which, in a one-phase region, yields

$$H = \frac{\partial \mathcal{F}_{MF}(\{M_j\}, \{Q_j\}, T)}{\partial M_i} \qquad \text{all } i,$$

$$\Delta = -\frac{\partial \mathcal{F}_{MF}(\{M_j\}, \{Q_j\}, T)}{\partial Q_i} \qquad \text{all } i. \tag{2.35}$$

In homogeneous phases for which all $M_i = M$, and all $Q_i = Q$, (2.35) reduces to two equations for the two unknowns $M(T, H, \Delta)$ and $Q(T, H, \Delta)$. These equations can have multiple solutions representing different phases. The phase diagram is obtained by standard methods, such as by comparing the Gibbs free energy of two different phases at the same values of the fields T, H and Δ. Continuous transitions can be obtained more easily by generating a Landau expansion. Such an expansion could be in terms of two order parameters, $M - M_0$ and $Q - Q_0$, where M_0 and Q_0 are the values corresponding to a single homogeneous phase and are obtained from (2.35). For the balanced system near the transition from a single phase to a coexistence of oil-rich and water-rich phases, $M_0 = 0$, and Q_0 is given by the solution of

$$\frac{Q_0}{2(1 - Q_0)} = \exp[(2dKQ_0 - \Delta)/T], \tag{2.36}$$

with d the dimensionality. The Landau expansion takes the form

$$\mathcal{F}_{MF}(M, Q, T) = \mathcal{F}_{MF}(0, Q_0, T)$$

$$+ \tilde{A}_2(T, Q_0)M^2 + \tilde{A}_4(T, Q_0)M^4 + \dots$$

$$+ \tilde{B}_2(T, Q_0)(Q - Q_0)^2 + \tilde{B}_3(T, Q_0)(Q - Q_0)^3$$

$$+ \tilde{B}_4(T, Q_0)(Q - Q_0)^4 + \dots$$

$$+ \tilde{C}_3(T, Q_0)M^2(Q - Q_0) + \tilde{C}_4(T, Q_0)M^2(Q - Q_0)^2 + \dots . \quad (2.37)$$

However, one knows that the order parameter which will become critical is M, not $Q - Q_0$. Therefore, one can generate a Landau expansion in terms of M alone. This is done by forming the Legendre transform of $\mathcal{F}(M, Q, T)$ with respect to Q only and then expanding this function in terms of M. Thus

$$\hat{\mathcal{F}}_{MF}(M, \Delta, T) \equiv \min_Q [\mathcal{F}_{MF}(M, Q, T) + N\Delta Q]$$

$$= \hat{\mathcal{F}}_{MF}(0, \Delta, T) + A_2(T, \Delta)M^2 + A_4(T, \Delta)M^4 + \dots . \quad (2.38)$$

where N is the number of lattice sites. The consolute line of continuous transitions, $T_c(\Delta)$, is determined by $A_2(T_c, \Delta) = 0$ as long as $A_4 > 0$. The location of the tricritical point, T_{tri}, and Δ_{tri}, is given by $A_2(T_{tri}, \Delta_{tri}) = A_4(T_{tri}, \Delta_{tri}) = 0$.

From the analysis of many different lattice models, the advantages and disadvantages of the mean-field approximation are well known. Mean-field theory ignores correlation effects, and thus underestimates the effect of thermal fluctuations. Such fluctuation effects become important in the vicinity of critical points, in particular in lower dimensions d. Thus, by comparing the two phase diagrams in Figs. 2.1 and 2.2, we can get an idea how important the effect of the fluctuations is for the kind of behaviour we are interested in. A brief glance at the two diagrams shows immediately that

(i) the phase transition temperatures are shifted by fluctuation effects,
(ii) fluctuations increase the region of the disordered phase in which it exists between the lamellar phase and oil–water coexistence,
(iii) the general structure of the phase diagram remains unchanged.

The three-dimensional phase diagram is as follows. At low temperatures and small values of Δ, there is two-phase coexistence between oil-rich and water-rich phases. If the temperature is raised, a consolute point $T(\Delta)$ is reached. The nature of the transition there is continuous for sufficiently small Δ, and is first order otherwise. A tricritical point, shown by an open circle, separates

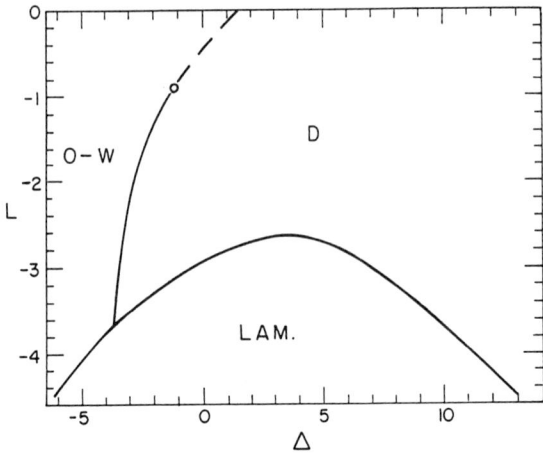

Fig. 2.3 Phase diagram of the three-component model at constant temperature, $T/J = 4.2$, and $K/J = 0.5$. The system is balanced and with equal concentrations of oil and water, $C = H = 0$. The phase diagram is shown as a function of the amphiphile strength L (in units of J) and the amphiphile chemical potential Δ (also in units of J). The region of disordered fluid is denoted D, that of the lamellar phase LAM, and that of oil-water coexistence O–W. From Lerczak *et al.* (1992).

these two regimes.[†] A lamellar phase exists over a region of amphiphile chemical potential. The transition to it is first order. The disordered phase coexists with oil-rich and water-rich phases down to a four-phase point. Below this point, the lamellar phase coexists with the water and oil phases.

One can make a more direct comparison with the experimental phase diagrams of the C_iE_j systems shown in Fig. 1.5 as follows. One notes from that figure that the temperatures at which the systems are balanced, (i.e. the temperatures of the tails of the fish), do not vary a great deal. As the number of head and tail groups increases, the strength of the amphiphilic interaction grows, which can be modelled by making the amphiphilic interaction L of (2.7) more negative. Therefore we show in Fig. 2.3 a phase diagram at constant temperature of the balanced system as a function of the amphiphilic strength L and the chemical potential of the amphiphile Δ. The same phase diagram shown as a function of amphiphile concentration is given in Fig. 2.4 (Lerczak *et al.*, 1992).

For weak amphiphiles, $|L|$ small, the transition from oil–water coexistence to the disordered phase is continuous. As the strength of the amphiphile is increased a tricritical point is passed, and the transition becomes first order so

---

[†]It has been argued by Matsen and Sullivan (1992b) that there should be no such tricritical point in two dimensions when the lamellar phase is present.

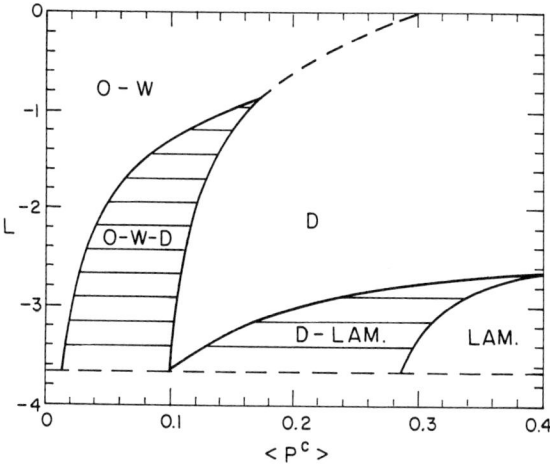

Fig. 2.4 The same phase diagram as in Fig. 2.3, but shown as a function of the amphiphile strength L (in units of J) and the amphiphile concentration $\langle P^c \rangle$. The region O–W–D is that of three-phase coexistence between oil, water and disordered fluid phases. Two-phase coexistence between lamellar and disordered fluid is denoted D–LAM. The dashed horizontal line near the bottom of the figure represents the value of L at four-phase coexistence. From Lerczak *et al.* (1992).

that there is three-phase coexistence between the oil-rich, water-rich and disordered (or middle) phases. There are no lamellar phases in the balanced system. These only appear when the amphiphile strength is increased even further. One sees that with the appearance of the lamellar phase, the region of concentration over which the middle phase exists decreases rapidly and ends at the four-phase point. We also see from Fig. 2.4 that the concentration of amphiphile in the oil and water phases decreases with increasing $|L|$ as does the concentration of amphiphile in the middle phase. Thus the strength of the interaction $|L|$ does correlate with the strength of the amphiphile as measured by its ability to solubilize oil and water. As the height of the three-phase triangle decreases, its base increases as the oil and water phases become purer, as shown in Fig. 2.5.

All of these results are in accord with the experimental results of Fig. 1.5. The main experimental feature which is not reproduced by the theoretical results is the monotonic decrease of amphiphile concentration in the middle phase as the number of head and tail groups is increased. In the model results, this concentration decreases initially, but then levels off as $|L|$ is further increased. This may be due to the use of mean-field theory for, as we shall discuss later in Section 4, thermodynamic fluctuations tend to keep the amphiphilic sheets apart and reduce the amount of amphiphile needed to

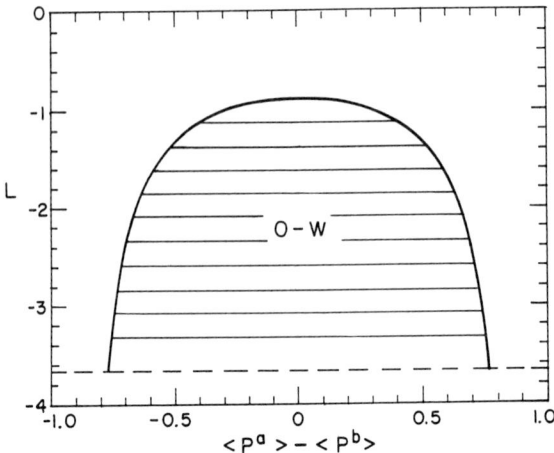

Fig. 2.5 The base of the three-phase triangle as a function of amphiphile strength (in units of J) is shown for the system whose phase diagrams are given in Figs. 2.3 and 2.4. From Lerczak *et al.* (1992).

solubilize the oil and water. The effect of these fluctuations is missed by the mean-field theory calculation.

A calculated phase diagram as a function of the three concentrations is shown in Fig. 2.6 for a balanced system not far from its four-phase point. It is interesting to compare it with the experimental phase diagram of the octane–water–$C_{10}E_5$ system shown in Fig. 1.7. Both diagrams show a three-phase triangle with a small height and wide base, so that the amphiphile is a good solubilizer and is rather insoluble in oil and water. Further, the middle phase is isolated from all other phases. One can easily see how with a small change of conditions, this middle phase will disappear entirely at a four-phase point. What the model calculation does not do well is to reproduce the lamellar phases near the binary sides of the phase prism. This is because, as noted earlier, the Hamiltonian of (2.7) does not contain a term which favours amphiphilic bilayers. Such a term, like that of (2.8), could easily be added, however.

2.2.2 Microemulsion structure

It is clear from Figs. 1.7 and 2.6 that the disordered phase is to be identified with the microemulsion (K. Chen *et al.*, 1987; Stockfisch and Wheeler, 1988; Gompper and Schick, 1989a) but this is not so clear from Fig. 2.4 in which we see that this disordered phase is continuously connected with the disordered phase occurring in the complete *absence* of an amphiphilic interaction, $L = 0$.

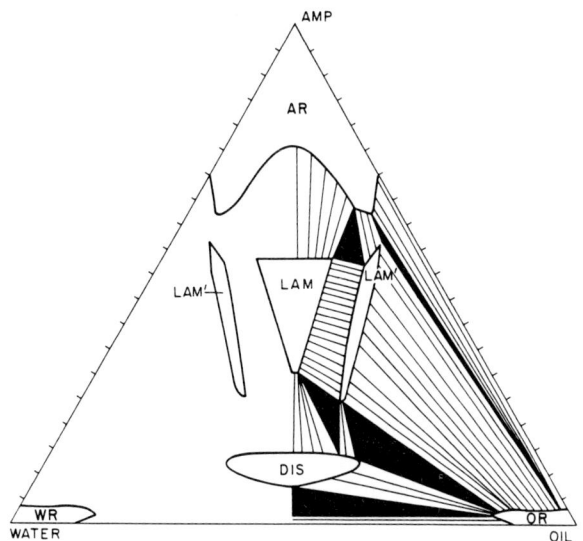

Fig. 2.6 Full phase diagram at $T/J = 4.45$ of the same system shown in Fig. 2.2. There are several different phases: an oil-rich (OR), a water-rich (WR), and a disordered fluid phase (DIS), all at low amphiphile concentration, a symmetric lamellar phase (LAM), two asymmetric lamellar phases (LAM') which are related by an interchange of oil and water, and an amphiphile-rich phase (AR). Three-phase coexistence triangles and tie lines (schematic) are shown on one half of the diagram. From Gompper and Schick (1990a).

What is the difference between these regions of the disordered phase? Presumably it is a matter of structure. In order to determine the structure of the fluid phase which coexists with the oil and water phases, correlation functions were calculated (Gompper and Schick, 1989a, 1990b) using transfer matrix methods (Nightingale, 1976) in two dimensions, and the Ornstein–Zernike approximation in three. In the transfer matrix approach, the correlation function $\langle M_0 M_r \rangle$ is calculated, where r is the distance along the infinite direction of the two-dimensional strip and M_i is the sum of spins in row i of finite length. From the mapping to the spin variables (2.3), this correlation function can be identified with $\langle (P_0^a - P_0^b)(P_r^a - P_r^b) \rangle$, the correlation function of the water–oil density difference. In terms of the eigenvalues λ_j and eigenfunctions $|j\rangle$ of the transfer matrix, the correlation function reads

$$\langle M_0 M_r \rangle = \sum_j \left(\frac{\lambda_j}{\lambda_0}\right)^{|r|} \langle 0|\hat{M}|j\rangle\langle j|\hat{M}|0\rangle \tag{2.39}$$

where $\hat{M}$ is the spin operator of a row of spins, and λ_0 denotes the largest eigenvalue. The asymptotic decay of the correlation function for large distances r is determined by the second-largest eigenvalue. In the system considered here, this eigenvalue can be *complex* (since the transfer matrix is not symmetric). Therefore, the correlation function has the general asymptotic form

$$\langle (P_0^a - P_0^b)(P_r^a - P_r^b) \rangle \propto e^{-r/\xi} \cos(qr + \phi), \qquad (2.40)$$

for large r, where ϕ is an unimportant phase factor. There are two length scales in this function: the correlation length ξ, which gives the typical size of correlated regions, and q^{-1}, which is the length over which significant changes in structure are found. In the microemulsion, we expect that q should be non-zero, for the oscillatory behaviour of the correlation function reflects the tendency of the amphiphile to order the oil and water. In an ordinary fluid, the wavevector $q = 0$, and the correlation function simply decays exponentially. In this sense, the fluid is structureless. It should be remembered, however, that the above correlation function does not include the structure factors of the molecules themselves which impart structure to the fluid at molecular lengths. But we are interested in structure at much larger distances, and it is this which is reflected in the above correlation function.

In the calculation, one finds that the fluid phase is characterized by a vanishing wavevector q in parts of the phase diagram, and by a non-vanishing q in others. The locus of points at which q just vanishes is called a disorder line (Fisher and Widom, 1969; Stephenson, 1970). In Fig. 2.7 we have reproduced the phase diagram of Fig. 2.3, and shown the disorder line as a dotted line; q is non-zero to the right of it. Note that over a large part of the

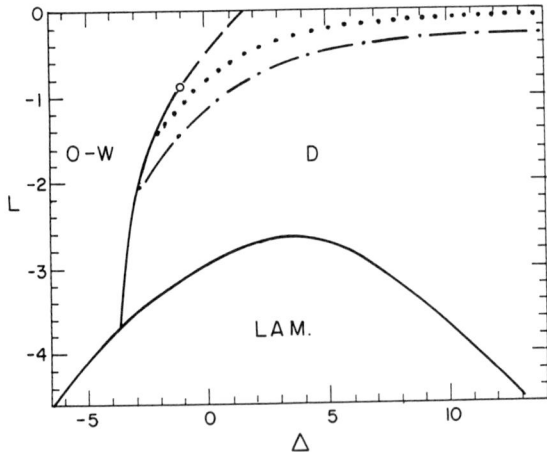

Fig. 2.7 The same phase diagram as in Fig. 2.3, but showing the disorder line (dotted) and Lifshitz line (dashed-dotted). From Lerczak *et al.* (1992).

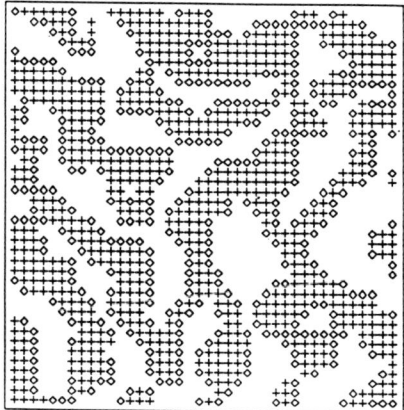

Fig. 2.8 A typical configuration of the microemulsion in the two-dimensional system with $K/J = 2$, $K_2/J = -0.2$, $L/J = -3$, at temperature $T/J = 1.1$ and chemical potential $\Delta/J = 5.2$, not very far from the transition to the lamellar phase. From Gompper and Schick (1990b).

triple line, the disordered phase is characterized by a non-zero value of q. The characteristic length, q^{-1}, is due to the same amphiphilic interaction which produces the relatively well-ordered lamellar phase at higher amphiphile concentrations, but which at smaller concentrations produces only a disordered phase of oil and water regions separated by amphiphile. *It is this phase, disordered but not structureless, which coexists with water-rich and oil-rich phases that we identify with the microemulsion.*

A typical configuration in the microemulsion phase, not very far from the transition to the lamellar phase, is shown in Fig. 2.8. The ordering of oil and water into microdomains, which are separated by sheets of amphiphiles, is clearly visible. Almost all amphiphiles are found to be located at the interface between oil and water regions, but not all oil–water pairs are separated by amphiphiles. Thus, in this model the sheets of amphiphile are not complete, but have holes in them.

It is not the correlation functions which are measured directly, but their Fourier transforms, or structure functions, which are determined in scattering experiments. Even though there are three components in the fluids, there are only two independent structure factors if the fluids are assumed to be incompressible. The individual water–water and amphiphile–amphiphile structure functions can be measured directly by neutron scattering with appropriate deuteration of one of the components, a process which exploits the very different neutron scattering lengths of hydrogen and deuterium (Auvray, 1994). The water–amphiphile structure function can then be

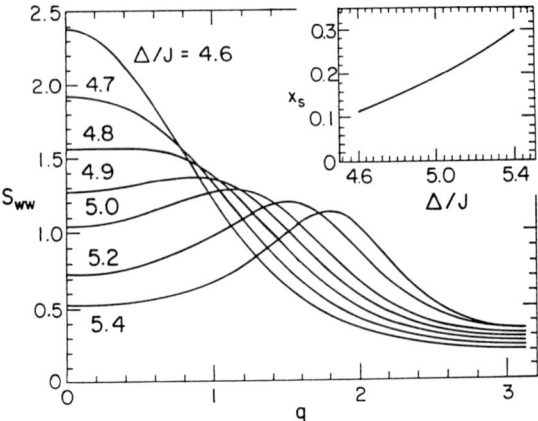

Fig. 2.9 Unnormalized water–water structure functions vs. wavevector in the (11) direction calculated at $T/J = 1.1$ for the system of Fig. 2.1. The wavevector is in units of $\sqrt{2}/a$ with a the lattice constant of the model. The values of Δ span the region of existence of the disordered phase at this temperature. The variation in amphiphile concentration with chemical potential Δ is shown in the inset. From Gompper and Schick (1989a).

inferred from a comparison of these results with the structure function of an undeuterated sample.

For a balanced system, the water–water structure function $S_{WW}(\mathbf{q})$ can be related to the Fourier transform of $\langle M_0 M_r \rangle$, denoted $\langle M_\mathbf{q} M_{-\mathbf{q}} \rangle$, and that of $\langle Q_0 Q_r \rangle$, denoted $\langle Q_\mathbf{q} Q_{-\mathbf{q}} \rangle$, according to $S_{WW}(\mathbf{q}) = [\langle M_\mathbf{q} M_{-\mathbf{q}} \rangle + \langle Q_\mathbf{q} Q_{-\mathbf{q}} \rangle]/4$, a relation which follows from the mapping between particle occupation variables and magnetic ones, (2.3). In the two-dimensional case, the two Fourier transforms can be calculated in terms of the eigenvalues of the transfer matrix (Bartelt and Einstein, 1986; Gompper and Schick, 1990b). A calculated water–water structure function in the two-dimensional disordered phase at $T/J = 1.1$ is shown in Fig. 2.9 in the range of Δ over which this phase exists. The variation of the amphiphile concentration, x_s, with Δ in this range is shown in the inset. It is approximately 10% at three-phase coexistence at this temperature. The wavevector $\mathbf{q}$, which is in the (11) direction, is in units of the inverse lattice constant of our lattice model; in these units, the basic reciprocal lattice vector is of magnitude 2π.

For the three-dimensional system the structure functions of the middle phase are calculated in the Ornstein–Zernike approximation as follows. We expand the mean-field free energy, (2.33), about the uniform values M_0 and Q_0 characterizing the middle phase. These values can be related to the fields H, T and Δ from (2.35), just as in the generation of the Landau expansion

(2.37). However, in generating that expansion, we wanted to know whether M_0 and Q_0 represented a single, stable phase, and therefore sought *uniform* deviations from these values $M - M_0$ and $Q - Q_0$ which would characterize other homogeneous symmetry-broken phases. Here, in contrast, we assume that the middle phase is stable, and consider the free energy cost of *nonuniform* deviations from it, for it is these fluctuations away from uniformity which are measured by the structure functions. Therefore we expand the local order parameters M_i and Q_i according to

$$M_i = M_0 + \sum_q M_q e^{i\mathbf{q}\cdot\mathbf{r}},$$

$$Q_i = Q_0 + \sum_q Q_q e^{i\mathbf{q}\cdot\mathbf{r}}, \tag{2.41}$$

and substitute these expressions into the mean-field free energy (2.33). This is then expanded to second order in the M_q and Q_q

$$\mathcal{F}_{MF}(\{M_i\}, \{Q_i\}, T) = \mathcal{F}_{MF}(M_0, Q_0, T)$$

$$+ N \sum_q [\alpha_q M_q M_{-q} + \beta_q Q_q Q_{-q}$$

$$+ \gamma_q (M_q Q_{-q} + M_{-q} Q_q)], \tag{2.42}$$

where the coefficients $\alpha_q(T, M_0, Q_0), \beta_q(T, M_0, Q_0)$, and $\gamma_q(T, M_0, Q_0)$ are known, and N is the number of lattice sites. The free energy is diagonalized by the rotation

$$M_q = \cos\theta_q X_q + \sin\theta_q Y_q,$$

$$Q_q = -\sin\theta_q X_q + \cos\theta_q Y_q,$$

$$\sin 2\theta_q = 2\gamma_q / R_q,$$

$$\cos 2\theta_q = (\beta_q - \alpha_q) / R_q,$$

$$R_q = [(\alpha_q - \beta_q)^2 + 4\gamma_q^2]^{1/2}. \tag{2.43}$$

It then takes the form

$$\mathcal{F}_{MF}(\{M_i\}, \{Q_i\}, T) = \mathcal{F}_{MF}(M_0, Q_0, T)$$

$$+ N \sum_q (A_q^- X_q X_{-q} + A_q^+ Y_q Y_{-q}), \tag{2.44}$$

where

$$A_q^{\pm} = (\alpha_q + \beta_q \pm R_q)/2. \tag{2.45}$$

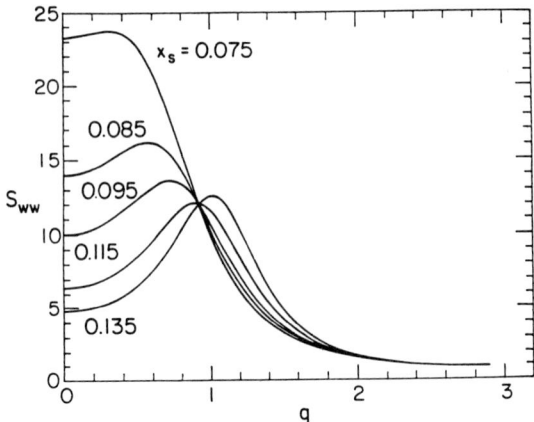

Fig. 2.10 Unnormalized water–water structure functions vs. wavevector in the (111) direction calculated at $T/J = 4.45$ for the system of Fig. 2.2. The wavevector is in units of $\sqrt{3}/a$. The amphiphile concentrations span the region of existence of the disordered phase at this temperature. From Gompper and Schick (1989a).

From this expression we immediately obtain two independent structure functions

$$S^-(\mathbf{q}) = \langle X_{\mathbf{q}} X_{-\mathbf{q}} \rangle,$$
$$= T/(2A_{\mathbf{q}}^-),$$
$$S^+(\mathbf{q}) = \langle Y_{\mathbf{q}} Y_{-\mathbf{q}} \rangle,$$
$$= T/(2A_{\mathbf{q}}^+). \tag{2.46}$$

Finally, from the relation between the molecular and magnetic variables, (2.3) and the linear transformation from the latter to the $X_{\mathbf{q}}$ and $Y_{\mathbf{q}}$ we relate the desired molecular structure functions to $S^-(\mathbf{q})$ and $S^+(\mathbf{q})$.

Figure 2.10 shows a result for the water–water structure function S_{WW} of the three-dimensional system at $T/J = 4.45$ for the range of amphiphile concentrations over which the phase exists. The amphiphile concentration in the disordered phase at three-phase coexistence is 7.5%.

For comparison, the intensity of neutron scattering from a toluene, brine, sodium dodecyl sulfate (SDS) microemulsion (Auvray *et al.*, 1984) is shown in the upper part of Fig. 2.11. This intensity is proportional to the water–water structure function. The similarities between the theoretical results and the experimental data are clear. In both cases, the peak in the water–water structure function moves out and decreases as the concentration of amphiphile increases. Furthermore, if one takes the size of the lattice constant to be 25 Å, a reasonable size for an amphiphile molecule, then the peaks in the

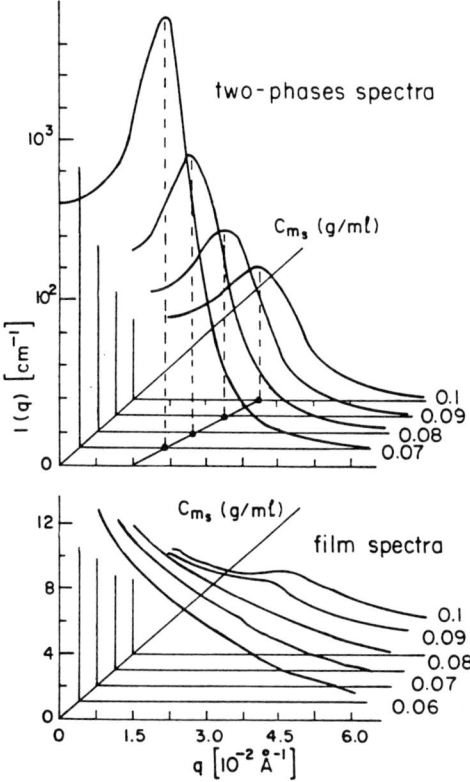

Fig. 2.11 Neutron scattering intensities from the system brine, toluene, and sodium dodecyl sulfate (SDS) as given by Auvray *et al.* (1984). The volume fraction of brine and oil are equal. The mass density of amphiphile is denoted C_{ms}. The upper spectra are proportional to the water–water structure function, the lower to the amphiphile–amphiphile structure function. (Reprinted with permission from L. Auvray, J.-P. Cotton, R. Ober and C. Taupin, *J. Phys. Chem.* **88**, 4586. Copyright 1984 American Chemical Society.)

calculated structure function, Fig. 2.10, occur at $3 \times 10^{-2} \text{Å}^{-1}$. This is a reasonable value as seen from the experimental data of Fig. 2.11. The agreement with this data should not be overemphasized as other systems show well defined peaks at much smaller wavenumbers. What is important is that the lattice theories, starting only with the cell size of molecular lengths, can *generate* characteristic sizes $2\pi/q$ which are much larger than a molecular length.

As the concentration of amphiphile decreases, the characteristic peak moves toward zero. We know from Fig. 2.7 that if the amphiphile concentration is decreased sufficiently, the disorder line will be crossed and the fluid will lose its structure, a circumstance which should be reflected in the structure function. In fact, as the disorder line is approached, but before it is reached, the peak in the structure function will go to zero wavevector. The locus at which this occurs is called the Lifshitz line (Hornreich *et al.*, 1975) and is, perhaps, a more experimentally useful delineator for the occurrence of a microemulsion. This line is shown in Fig. 2.7 as a dotted-dashed line. The difference between the disorder and the Lifshitz lines is that, at the former, the oscillatory behaviour first appears in the long distance behaviour of all correlation functions, whereas at the latter, the oscillatory behaviour first dominates the structure function. Clearly, more amphiphile is needed for the latter than the former, in agreement with Fig. 2.7. There are no singularities in the thermodynamic functions on crossing either line.

The intensity shown in the lower part of Fig. 2.11 is proportional to the amphiphile–amphiphile structure function. In general, it does not show the same features as the water–water structure function, but is characterized by a peak at zero wavenumber. A small shoulder at non-zero q may develop as the amphiphile concentration is increased. For comparison, the calculated amphiphile–amphiphile structure function, $S_{SS}(\mathbf{q})$, is shown in Fig. 2.12(a) for the same temperature and range of Δ employed in calculating the water–water structure function of Fig. 2.10. The function decreases monotonically with wavevector. If we break the oil–water symmetry by taking C in (2.7) to be non-zero, then we find that a small shoulder develops in this structure function as the amphiphile concentration is increased. An example is shown in Fig. 2.12(b). The difference between water and oil concentrations is fixed at -0.2 while the amphiphile concentration is increased. The water–water structure function at these same values has a peak at non-zero wavenumber indicating that the disordered fluid is a microemulsion.

Finally, we show in Fig. 2.13(a) an example of the calculated water–amphiphile structure function. The system parameters and temperature are the same as those of Fig. 2.10. However, in that figure, the oil/(oil + water) ratio was fixed at 0.5 and the amphiphile concentration was varied. In Fig. 2.13, the amphiphile concentration is fixed at 0.1, and the oil/(oil + water) ratio is varied. For comparison, experimental results (Auvray *et al.*, 1986) are shown in Fig. 2.13(b). The theoretical curves correctly reproduce the principal feature of the experimental data: there is a maximum in the structure function for volume ratios of oil/(oil + water) larger than 0.6, while there is a minimum for volume ratios less than 0.4.

2.2.3 Interfacial tension

The interfacial tension between two phases which coexist at the values of the fields H, Δ and T is easily calculated within mean-field theory. One assumes

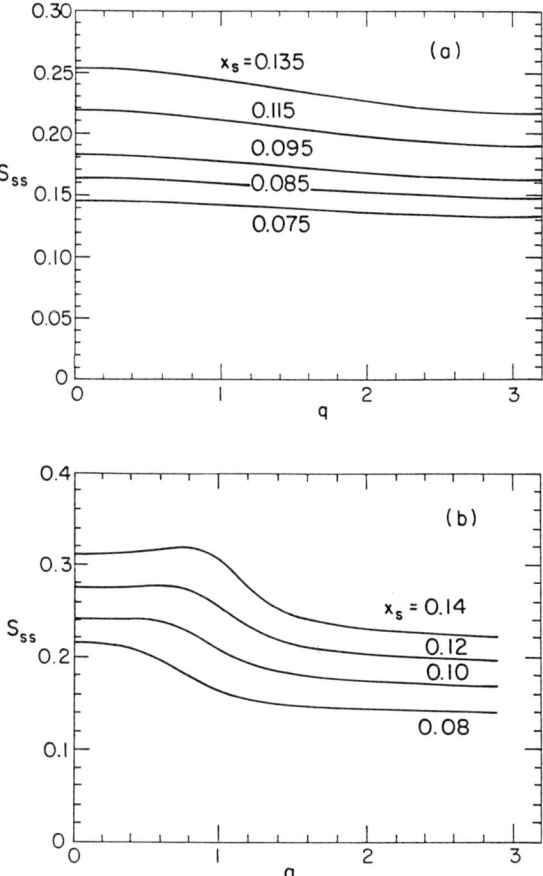

Fig. 2.12 (a) Amphiphile–amphiphile structure function for several amphiphile concentrations for the same parameters which apply to the water–water structure functions of Fig. 2.10; $T/J = 4.45$, $K/J = 0.5$, $L/J = -3.5$, $C/J = 0$, and equal concentrations of oil and water. The intensity scale is the same as in Fig. 2.10. (b) Amphiphile–amphiphile structure function for several amphiphile concentrations for a slightly different system; $T/J = 4.45$, $K/J = 0.5$, $L/J = -3.5$, but $C/J = 0.3$. The difference in volume fractions of water and oil is fixed at $\delta x = -0.2$. From Gompper and Schick (1990a).

that the densities M_i and Q_i vary in one direction only, and then solves (2.35) for some fixed number of planes, $p = 1, 2, ..., n$, which are normal to that direction. The solution must satisfy the boundary condition that the M_p and Q_p approach the values they take in the two bulk phases at $p = 1$ and $p = n$ respectively. Let these values be denoted $M^+(H, \Delta, T)$, $Q^+(H, \Delta, T)$ and $M^-(H, \Delta, T)$, $Q^-(H, \Delta, T)$. A series of approximations $\sigma_n(H, \Delta, T)$ to the

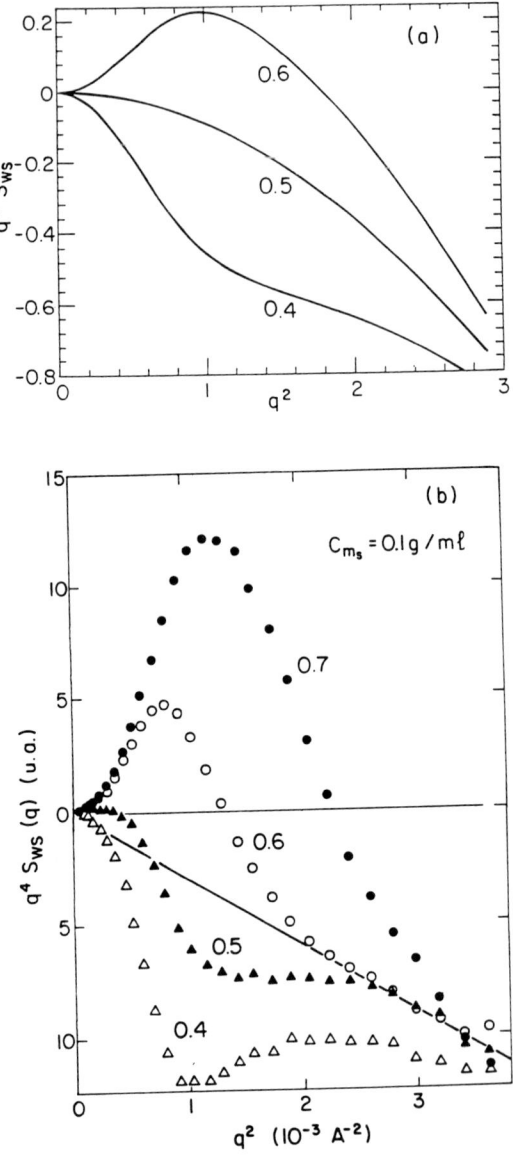

Fig. 2.13 (a) Scaled water–amphiphile structure function $q^4 S_{WS}$ vs. q^2 for three different volume ratios of oil/(oil + water). The system is the same as in Figs. 2.10 and 2.12(a). The amphiphile concentration is fixed at $x_s = 0.1$. From Gompper and Schick (1989a). (b) Scaled water–amphiphile structure function as determined by Auvray *et al.* (1986) for four different volume ratios of oil/(oil + water). The experimental system is the same as that in Fig. 2.11.

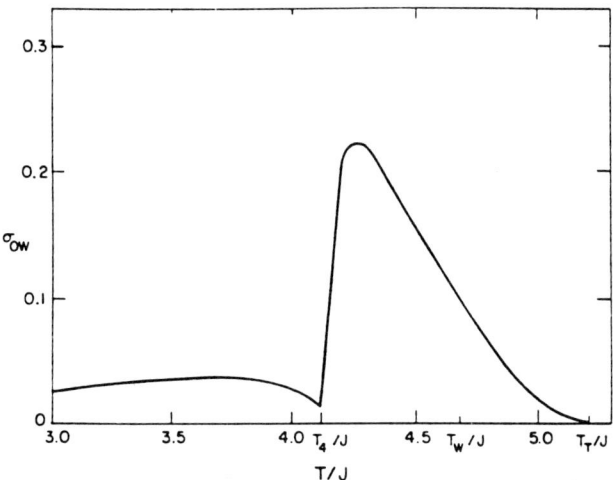

Fig. 2.14 Variation of the oil/water interfacial tension at three-phase coexistence. The four-phase coexistence temperature T_4, the wetting temperature T_W and the tricritical temperature T_T are indicated. The units of the interfacial tension are $J/\sqrt{3}a_0^2$. From Gompper and Schick (1990c).

interfacial tension $\sigma(H, \Delta, T)$ are obtained by calculating the difference of the Gibbs free energies (2.34) of the n-plane system with an interface and without. That is, we define g_{MF} to be the mean-field approximation to the Gibbs free-energy density at the plane p of the system with an interface by

$$g_{MF}(M_p, Q_p, T) \equiv f_{MF}(M_p, Q_p, T) - HM_p + \Delta Q_p, \qquad (2.47)$$

where f_{MF} is the mean-field approximation to the Helmholtz free-energy density. Similarly for the two bulk phases

$$g_{MF}(M^\pm, Q^\pm, T) = f_{MF}(M^\pm, Q^\pm, T) - HM^\pm + \Delta Q^\pm. \qquad (2.48)$$

Then

$$\sigma_n(H, \Delta, T) = \frac{1}{a_0^2} \sum_{p=1}^{n} \left\{ g_{MF}(M_p, Q_p, T) \right.$$

$$\left. - \frac{1}{2} \left[g_{MF}(M^+, Q^+, T) + g_{MF}(M^-, Q^-, T) \right] \right\}, \qquad (2.49)$$

where a_0 is the lattice constant. Finally, $\sigma = \lim_{n \to \infty} \sigma_n$. In Fig. 2.14 we show the oil/water interfacial tension along three-phase coexistence for the system of Fig. 2.2 in which the amphiphile strength L is constant and T is varied, and in Fig. 2.15 for the system of Fig. 2.3 in which T is fixed and L is varied. The interfacial tension at the temperature of the four-phase point of Fig 2.4,

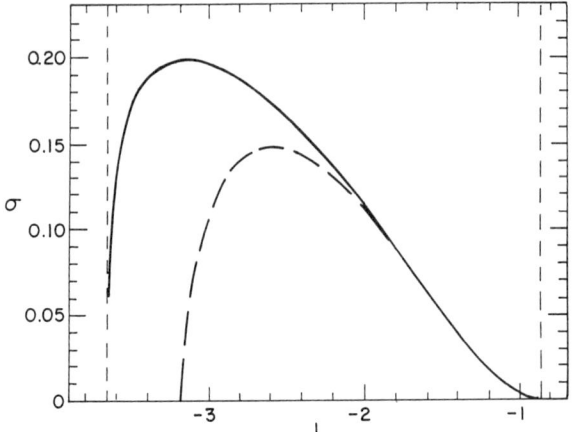

Fig. 2.15 Oil/water interfacial tension along three-phase coexistence with the disordered phase in a balanced system as a function of amphiphile strength L (in units of J). The units of the interfacial tension are $J/\sqrt{3}a_0^2$. The dashed line is obtained from the Ginzburg–Landau model which the lattice model generates (see Section 3.3.3 for details). From Lerczak *et al.* (1992).

$T_4/J \approx 4.1$, in the complete absence of amphiphile is about 1.7, so that the calculated reduction is of the order of 10^{-2}. The system of Fig. 2.15 is at a constant temperature $T/J = 4.2$ so again the interfacial tension of the oil–water system in the absence of amphiphile is approximately 1.7. The reduction of tensions here is a factor of 1/30. When fluctuations are included by simulating the model (Schmid and Schick, 1994), one obtains a reduction of 1/300 which is quite comparable to the good amphiphile C_6E_3 (Aratono and Kahlweit, 1991).

Let us try to understand why the tension decreases as the four-phase point is approached. To do so, we consider the disordered system as consisting of amphiphilic sheets separating regions essentially indistinguishable from those of the oil- and water-rich bulk phases. For simplicity, let the system be balanced, and the average distance between sheets be π/q. Assuming that the amphiphile sheets in the disordered phase are identical to that at an oil/water interface, we can write the Gibbs free-energy density of the disordered phase as

$$g_{dis} = \frac{1}{2}(g_{oil} + g_{water}) + \frac{q}{\pi}(\sigma_{ow} - |g_{int}(q)|), \qquad (2.50)$$

where g_{int}, the free energy of the interaction between sheets, has been assumed to be negative (attractive interaction). At three-phase coexistence $g_{dis} = g_{oil} = g_{water}$, so that

$$\sigma_{ow} = |g_{int}(q)|, \qquad (2.51)$$

where $g_{int}(q)$ is to be evaluated at the wavevector characterizing the disordered phase at three-phase coexistence. We expect the magnitude of g_{int} to decrease as the distance between amphiphilic sheets increases, i.e. as q decreases.

The same argument applies at three-phase coexistence between the lamellar and oil and water phases. At zero temperature in the model, the interaction between lamellae in their ground state configuration $+0-0+\ldots$ vanishes due to the very short range of the interactions. Thus, from (2.51), the interfacial tension is zero at zero temperature (Schick and Shih, 1986). The approach to this is seen already in Fig. 2.14. The vanishing of the tension makes the oil and water phases unstable to the lamellar phase. At non-zero temperatures, entropy effects make the tension between oil and water phases, in the presence of the lamellar phase, non-zero but small (Dawson, 1987). This result extends to the oil/water tension in the presence of the microemulsion, because the tension is continuous at the four-phase point. As the system is moved from the four-phase point, the interfacial tension increases before falling again to zero at the tricritical point.

An interesting correlation is illuminated by (2.51). A strong amphiphile is, by definition, a good solubilizer of oil and water, which means that the amount of amphiphile in the middle phase is small. Therefore, the sheets of amphiphile are rather far apart, and hence interact weakly. It follows from (2.51) that strong amphiphiles will produce low oil/water interfacial tensions, a correlation which is experimentally correct. (See, for example, Aratono and Kahlweit (1991).) This correlation also emerges from the three-component model as Figs. 2.4 and 2.15 show. From the former, one sees that the ability to solubilize oil and water is correlated with the strength of the amphiphilic interaction $|L|$, and from the latter that the interfacial tension decreases with this increasing strength.

The above mean-field calculation takes account of short-range forces on the interfacial tensions, but does not contain the effects of long-range van der Waals forces, or of thermodynamic fluctuations of the amphiphilic sheets. The former causes an attraction between sheets and therefore, according to (2.51), increases the interfacial tension. The latter brings about a repulsion of entropic origin (to be discussed further in Section 4.5) and thus decreases the tension. These forces will alter the specific relationship between the interfacial tension and the strength of the amphiphile, as measured by its ability to solubilize oil and water, but not the general property that the tension decreases as the strength increases.

2.2.4 Wetting properties

Finally, we examine results of the three-component model which are pertinent to the question of whether the middle phase should wet the oil/water

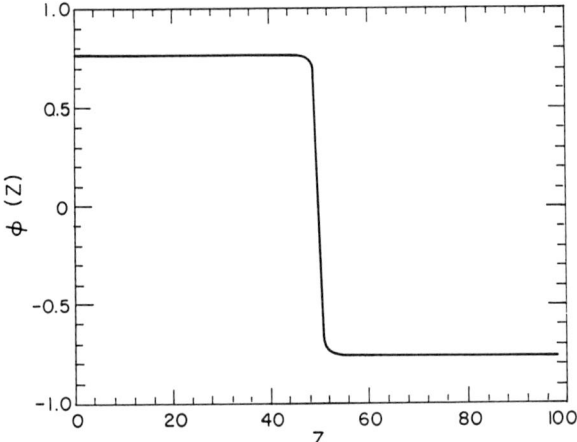

Fig. 2.16 Interfacial profile between water and oil phases for the system of Fig. 2.3 with an amphiphile of strength $L/J = -3.5$. The coordinate Z is measured in units of $a_0/\sqrt{3}$ where a_0 is the lattice spacing. From Lerczak *et al.* (1992).

interface or not. This can be answered in at least two ways. First, all three interfacial tensions, σ_{ow}, σ_{om} and σ_{wm} can be calculated in the manner indicated above. Then, the disordered phase does not wet the oil/water interface if

$$\sigma_{ow} < \sigma_{om} + \sigma_{mw}, \qquad \text{not wet} \qquad (2.52)$$

and does so if

$$\sigma_{ow} = \sigma_{om} + \sigma_{mw}, \qquad \text{wet.} \qquad (2.53)$$

An alternative method is to examine directly the interfacial profiles between oil and water phases calculated by the method above to determine whether there is a wetting film of middle phase between them. For example, a profile of the order parameter $\phi = \langle P_i^a - P_i^b \rangle$, the local difference between water and oil concentrations, is shown in Fig. 2.16 for a system which is near the four-phase point. As we have seen, near the four-phase point the middle phase which coexists with oil and water represents a microemulsion.

Clearly, between the oil and water phases, there is no wetting film of middle phase, for which the order parameter $\phi = 0$. In contrast, Fig. 2.17 shows a profile between oil and water phases for a system containing a much weaker amphiphile. That the middle phase wets the oil/water interface is clear.

One finds in the model that a wetting transition occurs when the disorder line is crossed. That is, if the disorder line is taken to be the dividing line between ordinary fluids and microemulsions, then microemulsions do not wet the oil/water interface while ordinary fluids do. The result that middle phases

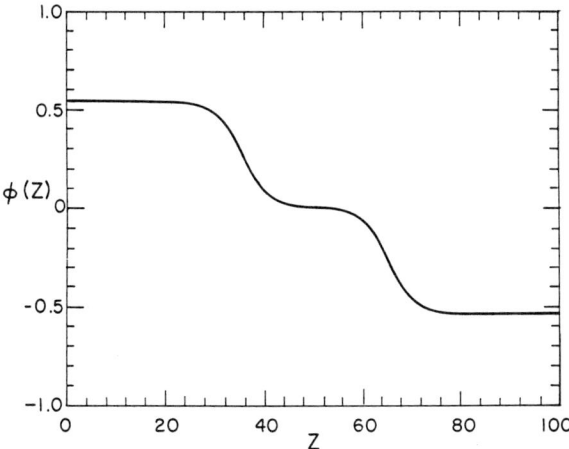

Fig. 2.17 Interfacial profile for the system of Fig. 2.3 with an amphiphile of strength $L/J = -1.45$. From Lerczak et al. (1992).

containing strong amphiphiles do not wet the oil/water interface while those containing weak ones do is in accord with experiment. The nature of the transition is unclear from the mean-field calculation, as is the reason why it occurs at the disorder line. Insight into this is provided by a Ginzburg–Landau theory, discussed in Section 3.3.2, which enables predictions to be made concerning the order of the transition and its location.

2.3 Some results of other lattice models

2.3.1 The Widom model

This model, with the Hamiltonian of (2.14), has been studied rather extensively, particularly by mean-field theory (Widom, 1986; Dawson, 1987; Dawson et al., 1988) and Monte Carlo simulation (Stauffer and Jan, 1987; Jan and Stauffer, 1988; Dawson et al., 1990; Stauffer and Eicke, 1992; Morawietz et al., 1992), as well as other methods (Levin and Dawson, 1990; Berera and Dawson, 1990; Kahng et al., 1990; Berera and Kahng, 1992), and has been extended to encompass additional interactions (Hofsäss and Kleinert, 1988; Hansen et al., 1991; Cappi et al., 1992). The calculated phase diagrams are usually given in the space of dimensionless coupling constants J/k_BT and M/k_BT, where k_B is Boltzmann's constant. In order to compare more easily with the phase diagrams of the three-component model, we show schematically, in Fig. 2.18, the phase diagram of the Widom

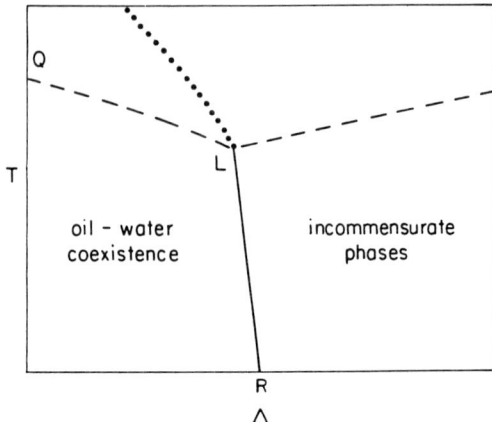

Fig. 2.18 Schematic phase diagram of the Widom model in three dimensions in the temperature and amphiphile chemical potential plane. The concentrations of oil and water are equal. Continuous transitions are shown by dashed lines, first-order transition by a solid line. The dotted line is a disorder line. The tricritical Lifshitz point is denoted L.

model in the space of temperature and amphiphile chemical potential. There are three distinct regions. At low amphiphile chemical potential, there is a coexistence region between oil-rich and water-rich phases. As the temperature is increased a consolute line, denoted QL in the figure, is reached. This is a line of continuous transitions to the disordered phase. If the amphiphile chemical potential is increased, a triple line, LR, is encountered at which the oil-rich, water-rich and "incommensurate" lamellar phases coexist. This region actually consists of many different phases, each separated from those adjoining it in phase space by first-order transitions. The particular phase which coexists with oil and water is periodic, the basic unit consisting of an oil region of thickness two lattice cells followed by a region of water of equal thickness. The oil and water are separated by amphiphile. The other phases differ from this one either in the thickness of oil or of water regions, or by being periodic in more than one direction. This region is quite similar to the corresponding one in the well-studied axial next-nearest-neighbour Ising (ANNNI) model (for reviews see Bak (1982); Fisher and Huse (1982); Selke (1988, 1992)).

The region of incommensurate phases was originally identified as the microemulsion by Widom (1986), but the more plausible suggestion that the disordered phase receive this identification was later made by Dawson *et al.* (1990). By extension, the incommensurate phases are identified as lamellar. The disordered phase in the Widom model is structured just as in the three-component model, and exhibits a structure factor with a peak at a non-zero wavevector (Widom, 1989; Dawson *et al.*, 1990; Levin *et al.*, 1992). Although the identification of the disordered phase with the microemulsion is probably correct, it does imply a difficulty with the model in that the disordered phase does not coexist with oil and water phases as the micro-emulsion does. Rather, the line of transitions to this phase is continuous. As a consequence, the oil/water interfacial tension is identically zero along this line, and the disordered phase just above this line is structureless having a structure function with a peak at zero wavevector. However, an extension of the range of interactions included in the model is sufficient to bring about a three-phase coexistence between microemulsion and oil-rich and water-rich phases (Hansen *et al.*, 1991).

The transition from the disordered phase to the incommensurate phases in the Widom model is first order (Stauffer and Jan, 1987; Jan and Stauffer, 1988; Dawson *et al.*, 1990). This is expected in a continuum model due to the fluctuations associated with the orientational degeneracy of the lamellar phases (Brazovskii, 1975; Dawson *et al.*, 1990; Levin and Dawson, 1990). That this expectation should also apply to a lattice model follows from the general assumption that, on large enough length scales, the underlying lattice structure of such a model will not affect the nature of phase transitions. When the nature of the phase transition is relatively simple, as in the continuous transition from the oil and water phases to the disordered phase, there is no reason to doubt that this is so. However, as the transition from the disordered to the lamellar phases is induced by fluctuations between orientationally equivalent phases, an equivalence which is not built into a lattice model, this assumption must be checked more carefully.

A minimal requirement that the orientational invariance be established is that the oil/water interface be rough, i.e. that the interface not be pinned by the underlying lattice. For a rough interface, fluctuations are controlled by the continuum capillary wave Hamiltonian (Weeks, 1980). Kahng *et al.* (1990) studied the roughness of oil/water interfaces within the Widom model. The results indicate that the roughening transition occurs at oil–water coexistence at temperatures *below* the transition to the disordered phase, in agreement with expectations (Weeks, 1980). The line of roughening transitions seems to pass through the point L, and to continue on the lamellar side of the disordered/lamellar phase transition. This indicates that results for the disordered to lamellar phase transition obtained from the lattice model are unaffected by lattice effects. However, near the transition to the lamellar phase, interactions between the interfaces become important; these inter-actions lead to a decrease of the interface roughness and may well induce a pinning of the monolayers to the lattice. More work is necessary to clarify this point.

If the transition to the lamellar phase were continuous, then the point denoted L in Fig. 2.18 would be a Lifshitz point (Hornreich *et al.*, 1975). However, as the transition is first order, it is probable that this point is a Lifshitz tricritical point (Aharony *et al.*, 1987), although an ordinary tricritical point cannot be excluded. Fig. 2.18 is drawn with the assumption that the former is correct.

2.3.2 The Alexander model

This model, with the Hamiltonian of (2.19), has been studied by mean-field theory and Monte Carlo simulation (K. Chen *et al.*, 1987, 1988; Stockfisch and Wheeler, 1988, 1993). The mean-field phase diagram for a two-dimensional system with a particular choice of parameters is shown in Fig. 2.19. The topology is probably very similar to the model's three-dimensional phase diagram, with one exception noted below.

Although the diagram is shown in Fig. 2.15 in the space of temperature and the strength of the amphiphilic interaction, the topology is presumably the same in the space of temperature and amphiphile chemical potential. There is a region of two-phase coexistence between oil-rich and water-rich phases. As the temperature is increased, the consolute line is reached. For weak amphiphiles, this transition to the disordered phase is continuous, while for stronger

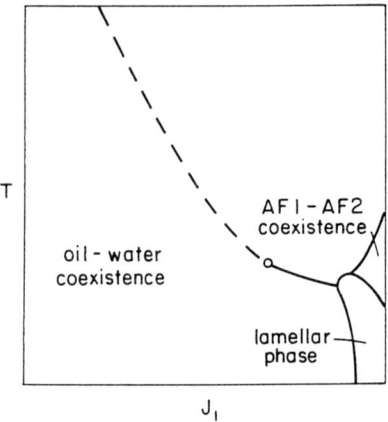

Fig. 2.19 Phase diagram of the Alexander model modified by K. Chen *et al.* (1987, 1988), as determined by mean-field theory for two dimensions. The system is symmetric, so that, in (2.19), $E^{aa} = E^{bb}$, $E^{aca} = E^{bcb}$, and $\mu^a = \mu^b$. In addition, the interaction between amphiphiles is $\tilde{E}^{cc} = 0$, and $E^{cc} = -0.2$. The diagram is shown in the space of temperature, measured in units of $(E^{aa} - E^{ab})/2$, and the strength of the three-particle interaction, $J_1 \equiv (E^{aca} - E^{acb})/2$. A continuous transition is shown by the dashed line, first-order transitions by solid lines. The tricritical point is shown by an open circle.

amphiphiles it is first order. A tricritical point separates the two regimes. As K. Chen *et al.* (1987, 1988) identified the disordered phase with the microemulsion, they obtained three-phase coexistence between microemulsion, oil and water. As the amphiphile strength is increased at low temperatures, the third phase in coexistence with oil and water is no longer disordered, but is a layered phase consisting of single sheets of oil, amphiphile, water, amphiphile, oil, etc. Presumably, this should be identified with the lamellar phase. Another ordered phase with oil and water in a checkerboard pattern, the two possibilities in coexistence, also appears and is denoted AF1 and AF2. (In three dimensions, an ordered phase with different structure probably appears here.) The phase behaviour leads to low interfacial tensions just as in the three-component model; the oil/water interfacial tension vanishes at the high-temperature end of the triple line, the tricritical point, and at the zero-temperature end also. Thus it is not expected to be large anywhere.

One difficulty with this model in the version studied is that it produces no long-period lamellar phases, i.e. ones with period longer than twice the basic lattice spacing. A consequence for the microemulsion is that it shows no structure at a long length scale. This is reflected in the structure function which has only a peak at zero wavevector, in contrast to experimental results. Thus the distinction is lost between the disordered phase produced by a strong amphiphile, which is a microemulsion, and that produced by a weak amphiphile, which is not. This problem is related to the short range of interactions taken; it is not difficult to conceive of only slightly longer ranged ones which would produce the long-period lamellar phases and also a structured microemulsion.

The interfacial tension between oil and middle phases at three-phase coexistence has been calculated by Stockfisch and Wheeler (1993) by simulating the model. At the point on the triple line at which the simulation is carried out, they find the oil/middle phase tension lower than a bare oil/water tension by a factor of about $6 \cdot 10^{-4}$. As the oil/water tension on the triple line is approximately twice the oil/middle phase tension (i.e. the middle phase is not very far from wetting the oil/water interface), the reduction in oil/water tension is about 1/800, which is stronger than the good amphiphile C_6E_3 (Aratono and Kahlweit, 1991).

2.3.3 Vector models

The most-studied class of vector models is that in which only the amphiphile is described by a vector. The representative Hamiltonian for a discrete-vector system interacting with nearest neighbours only and placed upon a cubic lattice is that of Matsen and Sullivan (1990) and is given in (2.25) and (2.26). This model has been studied by mean-field theory (Ciach *et al.*, 1988, 1989,

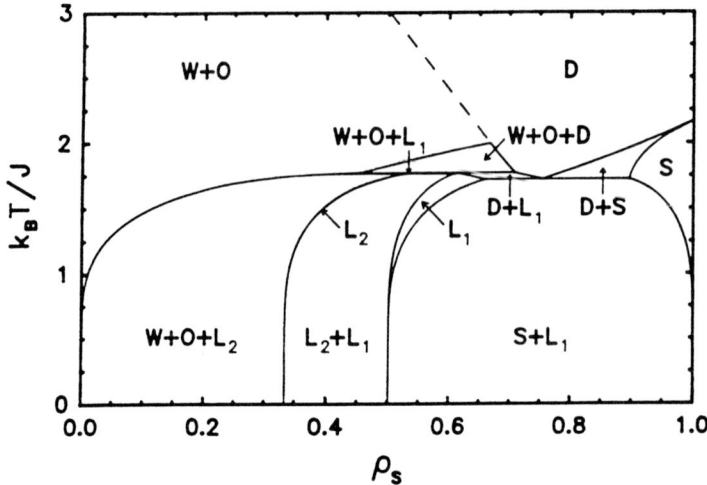

Fig. 2.20 Phase diagram for a ternary mixture in the temperature and amphiphile density plane, as calculated within mean-field theory for the vector model (2.25), (2.26) with discrete spins. The interaction constants are $J_1 = J_2 = \frac{4}{3}J_3 = \frac{4}{5}J_4 \equiv J$, and $K_i \equiv 0$, $i = 1, ..., 4$. The solid lines denote first-order transitions, and the dashed line denotes a second-order transition. The water-rich, oil-rich, and disordered phases are denoted W, O and D, respectively. The phase S is comprised of bilayers of amphiphile, and therefore has period two. In the lamellar phases L_1 and L_2 a single or double layer of oil and water appears between the layers of amphiphile; these phases have period four and six, respectively. From Matsen and Sullivan (1990).

1991; Ciach and Høye, 1990; Ciach, 1990, 1992; Matsen and Sullivan, 1990, 1992a), Bethe approximation (Matsen and Sullivan, 1990, 1992a), and Monte Carlo simulations (Laradji et al., 1991a; Gunn and Dawson, 1992). Versions appropriate to lower dimensions have been studied by transfer matrix methods in one (Matsen and Sullivan, 1991; Renlie et al., 1991) and two dimensions (Matsen and Sullivan, 1992b). The phase diagram for the three-dimensional system as obtained from mean-field theory is shown in Fig. 2.20. The general structure is similar to the results of the three-component model, except that there is now a transition to a liquid crystalline phase S at very high amphiphile concentration, a phase which the lattice models with scalar amphiphiles cannot describe. The phase diagram reveals that the amphiphile concentration in the microemulsion at three-phase coexistence with oil and water is high, about 65%, much larger than in the three-component model. Thus this calculation describes a weak amphiphile. In accordance with this, the progression from oil–water coexistence through a one-phase microemulsion region to the lamellar phase, typical of a strong amphiphile, does not exist. Neither of these effects are due to inherent problems of vector models.

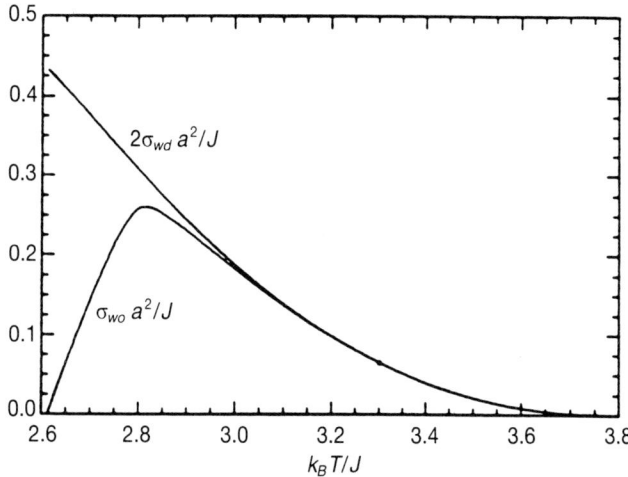

Fig. 2.21 A plot of the interfacial tensions between water and oil, and between water and disordered phases. The lattice parameter is a. From Matsen and Sullivan (1992a).

That the calculation results in the description of a weak amphiphile is probably due to the mean-field approximation which was used as it ignores the strong multi-particle correlations induced by the amphiphile as well as the effect of fluctuations. This is also true of the Bethe approximation, also employed by Matsen and Sullivan (1992a).

The same authors have also calculated the interfacial properties of this model. There are several noteworthy features. The interfacial tension σ_{ow} between oil and water and σ_{wm} between water and the disordered fluid are shown for the balanced system in Fig. 2.21 from the tricritical point at $k_B T/J \approx 3.77$ down to the four-phase point $k_B T/J \approx 2.61$. At temperatures above $k_B T_W/J = 3.30$, the oil/water interface is wetted by the middle phase so that $\sigma_{ow} = 2\sigma_{wm}$. Below this temperature, the interface is not wet. The wetting transition occurs very near the disorder line, and is weakly first order. The oil/water tension falls rapidly as the four-phase point is approached, and at that point it is reduced by a factor of $1/500$ from its value in the absence of amphiphile.

Interfacial profiles have been calculated, and that between the disordered and the water-rich phase is shown in Fig. 2.22. The density of oil is given by $p_\bullet$, that of water by p_o, the density of amphiphiles with heads pointing toward the water by $p_\rightarrow$, with heads pointing away from the water by $p_\leftarrow$, and with heads parallel to the interface by $p_\uparrow$. Note the small oscillations in the various densities on the disordered side (Gompper and Schick, 1990c; Chowdhury

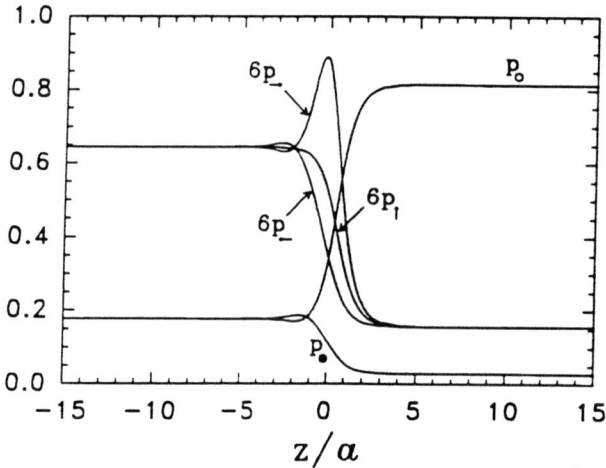

Fig. 2.22 Density profiles across the interface between water-rich and disordered phases at $k_B T/J = 2.8$ as calculated in mean-field theory. From Matsen and Sullivan (1992a).

and Stauffer, 1991). These reflect the bulk correlation function of the middle phase which, being on the microemulsion side of the disorder line, is an exponentially decaying oscillatory function.

An interesting result of Matsen and Sullivan (1992a) is the prediction of several phase transitions within the interface itself. At oil–water coexistence with little amphiphile in the system, the interface is bare. On adding amphiphile at low temperature, a first-order surface layer transition occurs and the interface becomes covered by a single layer of amphiphile. At higher temperatures, several layers of amphiphile condense simultaneously. These two first-order transitions meet at a triple point. From this point emerges a line of first-order transitions between the thick and thin interfaces, a line which ends at a critical point. These transitions are analogous to those predicted for adsorbed systems (Pandit et al., 1982).

Another phase diagram with an unusual feature has been obtained recently from the same vector model (Ciach et al., 1991; Ciach, 1992), with $J_2/J_1 = 3$, $J_1 = K_1$. A new phase appears, which resembles an ordered bicontinuous phase, with a coordination number four in the oil and water sublattices. The distance between neighbouring water molecules is equal to the diagonal of an elementary cell of the underlying simple cubic lattice. This phase can coexist with a disordered one, and it is tempting to identify the microemulsion with the latter. It should be noted, however, that the concentration of amphiphile in this microemulsion is very large, about 70%, and that the progression with increasing amphiphile concentration from oil–

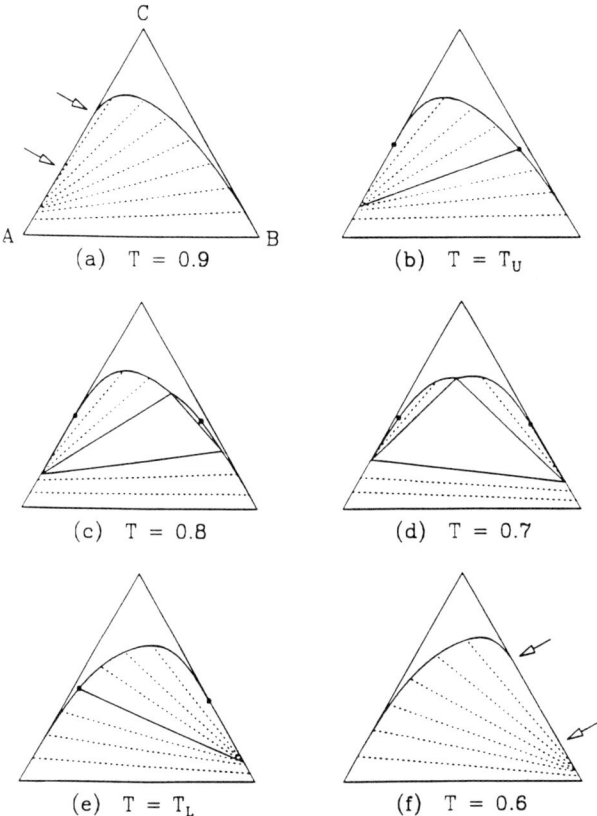

Fig. 2.23 A sequence of six isothermal phase diagrams for the ternary system of A (water), B (oil) and C (amphiphile). Critical points, tie lines and critical tie lines are shown with dots, dashed lines and solid lines respectively. The arrows in (a) and (f) mark the limits of the two-phase region along the A-C and B-C sides respectively. From Matsen *et al.* (1993).

water coexistence through a one-phase microemulsion region, to an ordered (lamellar of bicontinuous) phase is missing. Thus, this calculation also describes a weak amphiphilic system.

Certainly the best agreement with experiments on non-ionic amphiphilic systems results when the orientational properties of the water as well as those of the amphiphile are taken into account. This was done by Matsen *et al.* (1993) by means of the Hamiltonian of (2.30) and (2.31). The properties of the system were obtained from the Bethe approximation. Figure 2.23 gives the phase diagram as a function of temperature. It shows the experimentally observed progression with temperature from two- to three- to two-phase

G. Gompper and M. Schick

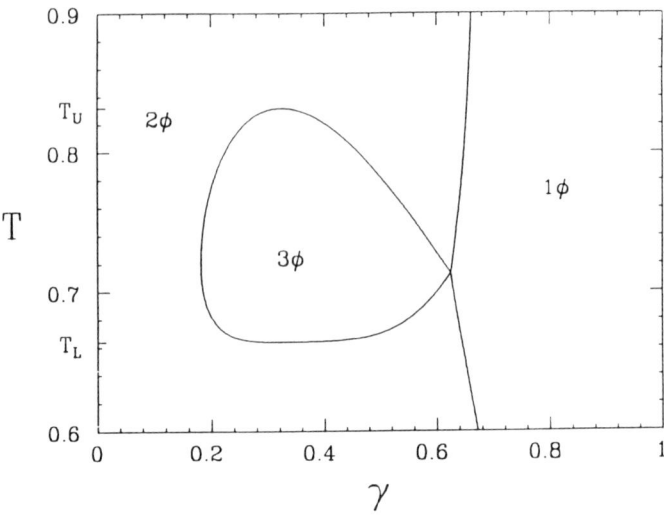

Fig. 2.24 A vertical slice through the phase prism for equal oil and water concentrations. The three-phase region exists in the temperature interval T_L to T_U. From Matsen *et al.* (1993).

coexistence. A cut at constant water to oil ratio of 1:1 shows the "fish" in Fig. 2.24. A comparison with Fig. 1.5 shows that this is a good description of the weak amphiphile C_4E_1.

The concentration of micelles in amphiphilic systems is of great interest, but is infrequently calculated. To do so, one must employ a theory which is applicable near, and at, the limit of the binary water–amphiphile system or oil–amphiphile system because it is here that micelles are expected to form. Further one must define what the critical micelle concentration (cmc) shall be taken to be. Such a definition may be formulated in terms of the effect of micelle formation on some property, such as the interfacial tension, a definition which serves only in a regime of phase coexistence, or the osmotic compressibility (Wenzel *et al.*, 1989). Alternatively, the definition can be formulated in terms of an abrupt increase in the number of micelles, a number which can either be directly counted in a simulation, or calculated analytically within some approximation (Goldstein, 1986). The latter definition assumes, of course, a choice of just what configurations of amphiphile will be accepted as comprising a micelle. Matsen *et al.* (1993) employed a face-centred cubic lattice, in which each site has twelve nearest-neighbour sites. They defined a micelle in the water-rich (oil-rich) system as a cluster consisting of a central site, occupied either by oil (water) or amphiphile, and its twelve nearest-neighbour sites all of which are occupied by amphiphiles.

Further, the orientations of the amphiphiles, which by definition of the model are restricted to the nearest-neighbour directions, must be such that all tails (heads) point either to the central site or to other members of the cluster. The micelle concentration was then calculated within the Bethe approximation, and the cmc defined in terms of a rapid increase in this concentration. In this way, it was determined that the phases which become critical at the lower and at the upper critical end points are micellar solutions; i.e. their concentrations of amphiphile are beyond the cmc. In the coexisting non-critical spectator phase, however, the concentration of amphiphile is less than the cmc. Hence, they are are molecularly disperse. This is in agreement with experiment (Kahlweit *et al.*, 1990).

As the temperature is increased from the lower critical endpoint, the composition of the middle phase coexisting with oil- and water-rich phases changes; it is characterized by an increasing fraction of oil. One expects that at a certain temperature the oil will span the system; as the water already does so, the middle-phase will become bicontinuous (Scriven, 1976, 1977). Because the amphiphile should now form sheets which separate these two bicontinuous regions, one expects the number of micelles to fall precipitously. The micelle concentration as calculated in the Bethe approximation does, in fact, show a steep decrease as the temperature is increased from the lower critical endpoint. This can be seen from Fig. 2.25 where the micelle concentration, ρ, is plotted. It is normalized by the micelle concentration ρ_0 which would exist solely by chance in the model if only amphiphiles were present. In order to determine whether this decrease can be associated with the onset of bicontinuity, one must locate this onset.

The determination of the connectivity of the oil and water regions is a correlated percolation problem (Blossey and Schick, 1991; Skaf and Stell, 1992b, 1993). It is a percolation problem because one asks whether there are paths within the water and oil regions which span the system, and it is a *correlated* percolation problem because the probability of finding a water or oil molecule next to another is not random, as it is in a normal percolation problem, but rather is determined by Boltzmann probabilities which depend upon the interactions within the system. In a simulation, the connectivity of the oil and/or water regions can be determined directly (Stauffer and Eike, 1992) using well-known algorithms (Stauffer, 1985). The connectivity can also be determined analytically as shown as follows for the oil regions.

One adds to each oil molecule a Potts label λ, which takes the values from 1 to q (Murata, 1979). These Potts degrees of freedom interact via the Hamiltonian

$$\mathcal{H}_p = -E_p^{bb} \sum_{\langle ij \rangle} P_i^b P_j^b [\delta(\lambda_i, \lambda_j) - 1] - \mu_p^b \sum_i P_i^b [\delta(\lambda_i, 1) - 1], \qquad (2.54)$$

where μ_p^b is a chemical potential which favours the Potts value $\lambda = 1$. The

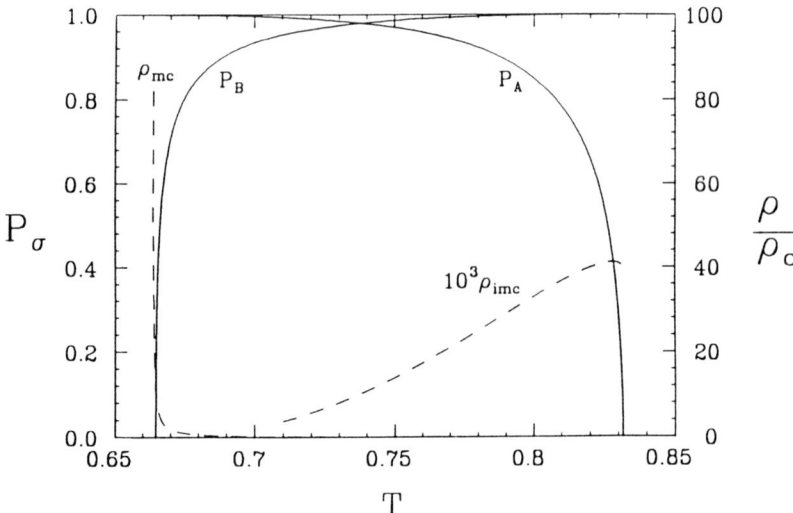

Fig. 2.25 Evolution with temperature of the middle phase in three-phase coexistence. Concentrations of micelles and inverse micelles are shown dashed. The probability P_A that a water molecule belongs to an infinite cluster, and P_B that an oil belongs to such a cluster, are shown with solid lines. From the beginning of the triple line at $T_L = 0.6639$ until the temperature at which $P_B = 0^+$, the middle phase is micellar. Between $P_A = 0^+$ and $P_B = 0^+$, it is bicontinuous. From the temperature at which $P_A = 0^+$ to $T_U = 0.8318$ at which the triple line ends, the phase is inverse micellar. From Matsen *et al.* (1993).

total Hamiltonian of the system is

$$\mathcal{H} = \mathcal{H}_0 + \mathcal{H}_{amp} + \mathcal{H}_p, \tag{2.55}$$

with the three pieces given by (2.1), (2.31) and (2.54) respectively. If there are oil molecules at sites i and j, and if the molecule at site i is in the favoured Potts state, then the effect of the first term in (2.54) is to make it more likely that the molecule at site j will also be in this state. It is this kind of correlation between nearest-neighbour sites which can be captured by the Bethe approximation. As Matsen *et al.* (1993) employed this approximation to determine the phase behaviour, it was convenient to employ it at the same time for the calculation of the connectivity properties.

The partition function of this system can be written in the following form (Kasteleyn and Fortuin, 1969; Wu, 1982)

$$Z = \text{Tr}\Big\{ \exp[-\beta(\mathcal{H}_0 + \mathcal{H}_{amp})] \, q^{N - \sum_i P_i^b} \sum_{\{\gamma\}} p^{R_\gamma}(1 - p)^{L_\gamma}$$

$$\prod_{G(\gamma)} [1 + (q - 1) \exp -(\beta \mu_p^b n_c)] \Big\}. \tag{2.56}$$

Here N is the total number of sites, and $p \equiv 1 - \exp(-\beta E_p^{bb})$ is the probability of a bond existing between sites occupied by oil; $\{\gamma\}$ is the set of ways one can introduce nearest-neighbour bonds on the restricted lattice consisting only of those sites occupied by oil; $G(\gamma)$ is the set of clusters present in one particular configuration γ; R_γ and L_γ are the number of bonds present and absent from this restricted lattice; n_c is the number of sites in the cth cluster in $G(\gamma)$. If the percolation cluster is defined such that it is sufficient that there be a connected path between nearest-neighbour sites occupied by oil, then this requires that the bond between occupied sites always be present. Hence one takes the limit $E_p^{bb} \to \infty$ before any other limits are taken. Then only the clusters with the bonds present between occupied oil sites contribute to the partition function. If an infinite spanning cluster exists, it will be characterized by the Potts state favoured by the applied chemical potential μ_p^b. One determines whether it exists or not by comparing the fraction of oil molecules in the non-favoured Potts states with the total fraction of oil molecules on the lattice. The fraction of sites which are in clusters characterized by any of the $q - 1$ Potts states not favoured by the chemical potential μ_p^b is obtained by partial differentiation of the partition function with respect to this chemical potential. If the limit $q \to 1$ followed by $\mu_p^b \to 0^+$ is taken, then one obtains the expectation value, $\langle P^b \rangle_{nf}$, of the fraction of oil sites which are not in the favoured cluster, the expectation value being taken in the ensemble governed by $\mathcal{H}_0 + \mathcal{H}_{amp}$ as desired for the physical problem

$$\langle P^b \rangle_{nf} = -\lim \frac{1}{N} \frac{1}{q-1} \frac{\partial \ln Z}{\partial \beta \mu_p^b}. \qquad (2.57)$$

The fraction of sites which are occupied by all oil molecules, irrespective of connectivity, is simply

$$\langle P^b \rangle_{all} = \frac{1}{N} \frac{\partial \ln Z}{\partial \beta \mu_p^b}. \qquad (2.58)$$

Thus, the fraction of oil molecules in an infinite spanning cluster is $P_B \equiv \langle P^b \rangle_{all} - \langle P^b \rangle_{nf}$. This probability is shown in Fig. 2.25 together with the concentration of micelles. One sees that, as expected, the concentration of micelles does fall precipitously just as the oil becomes continuous. Even though the result is expected, it is gratifying none the less, because it emerges from the comparison of two independent calculations. It therefore reinforces the belief that the physics of the system is being captured. One also sees from Fig. 2.25 that near the upper critical endpoint, the fraction P_A of water molecules in an infinite, spanning cluster falls rapidly and the number of inverted micelles grows. The number of inverted micelles near the upper critical endpoint is much less than the number of normal micelles near the lower critical endpoint because the interaction between oil and amphiphile is much weaker than between water and amphiphile.

The region of phase space in which the middle phase is bicontinuous would be probed experimentally by conductivity or self-diffusion measurements. Scattering experiments do not respond to the bicontinuity, of course, but probe other aspects of the structure of the middle phase. Markers for the onset of structure, the disorder lines, were calculated by Matsen *et al.* (1993). The region of phase space in which the water–water correlation function of the middle phase was characterized by an exponentially decaying oscillatory function indicating structure, and the region in which the middle phase was bicontinuous overlapped significantly, but were not identical. This only serves to reinforce the idea that the microemulsion is not a distinct thermodynamic phase, but is rather a disordered phase with certain experimental responses that serve to *define* it. These experimental responses may be those of scattering, conductivity or other measurements, but the choice is not unique.

2.3.4 The Larson model

In this lattice model, the amphiphile is permitted more internal degrees of freedom than in the models described above. It consists of i head units and j tail units (denoted $H_i T_j$); each head and tail unit is constrained to adjoin the neighbour, or neighbours, to which it is attached. Unit A is considered to adjoin unit B if A occupies one of the 26 sites in the $3 \times 3 \times 3$ cube of cells surrounding B. Thus the amphiphile is treated like a small polymer. Each head unit is identical to water in its interactions, and each tail unit is identical to oil in its interactions. Hence there is only one interaction parameter in the problem, $w \equiv (E^{aa} + E^{bb} - 2E^{ab})/2k_B T$. The model has been simulated extensively by Larson (1988, 1989, 1992). A phase diagram conjectured by him consistent with his results is shown in Fig. 2.26 for a temperature which is about 0.6 of the value of the critical point on the oil–water side. The amphiphile is $H_4 T_4$. One sees that at this temperature, the system is below its four-phase point as three-phase coexistence is between oil, water and lamellar phases. The concentration of amphiphile in the latter is somewhat beyond 0.33, the value it would attain at zero temperature.

The transition to the lamellar phase is interesting in that it appears to proceed quite differently for large and small amphiphile concentrations. For a concentration of 0.6, the ordering transition is quite sharp, whereas at 0.2, a concentration within the three-phase triangle, it seems to take place in several stages. First the amphiphiles assemble into sheets separating oil and water domains. The sheets bend on small length scales, and the tails of the amphiphile are also bent. As the temperature is lowered, the amphiphile chains stretch, and the sheets flatten into imperfect lamellae punctuated by holes, as

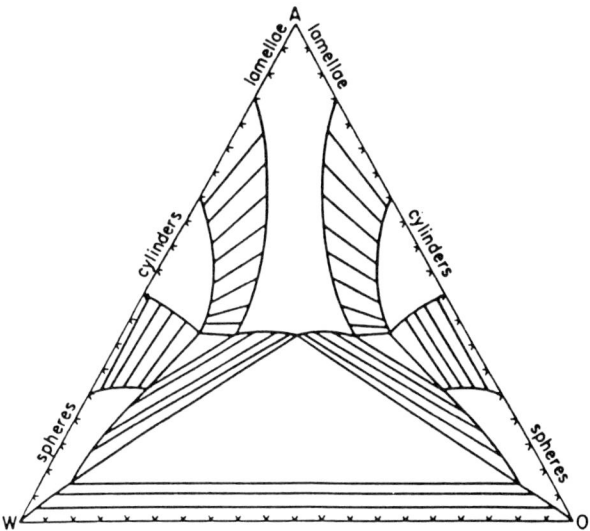

Fig. 2.26 Schematic phase diagram of oil, water, H_4T_4 for $w = 0.1538$. From Larson (1989).

shown in Fig. 2.27. With further cooling, the holes disappear. The stretching of the chains as the temperature is lowered, found also in simulations of longer chains (Minchau *et al.*, 1990), had previously been observed in experiments on diblock copolymers (Hadziioannou and Skoulios, 1982; Almdal *et al.*, 1990). Perhaps the phase in which the lamellae have many holes should be identified with the microemulsion, but this will take additional simulations in order to determine just where the disordered phase exists.

One of the nice features of this model is that the number of head and tail units can be varied. Larson (1989) finds that neither H_1T_1 nor H_2T_2 form lamellar phases, while H_3T_3 and H_4T_4 do. This is in qualitative agreement with the experimental observations, shown in Fig. 1.5, that C_iE_i does not produce lamellar phases in the balanced system for small values of i.

2.4 Mixtures of water and amphiphile

The simplest system which displays self-assembly is a two-component mixture of water and amphiphile. The same kind of lattice models which have been applied to ternary mixtures can also describe the binary ones. Preferably, the latter should simply be a limit of the former. There are, again, several levels of complexity depending on the degrees of freedom

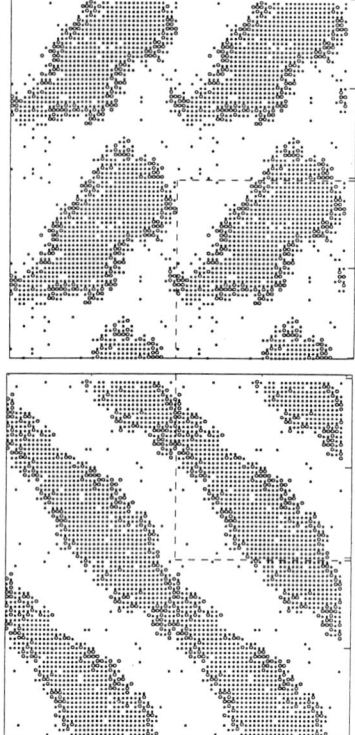

Fig. 2.27 Two mutually perpendicular slices from the structure formed by cooling to $w = 0.1385$ a system consisting of a concentration 0.2 of H_4T_4, and 0.4 each of oil and water. The lattice is $40 \times 40 \times 40$. Each of the slices are made from four repetitions of the basic unit shown enclosed in dotted lines. The asterisks are oil units, the circles are tail units; other units are not shown. From Larson (1992).

one attributes to the amphiphile and to the water. Models like those of (2.22) or of (2.26), in which the amphiphiles alone possess orientational degrees of freedom, degrees which are discrete or continuous, have been studied within mean-field theory by Gompper and Schick (1989b) and Matsen and Sullivan (1990), and within Bethe approximation by the latter. From these calculations, the following behaviour emerges. For small directional interactions, there is at low temperatures a two-phase coexistence between homogeneous water-rich and amphiphile-rich phases. At higher temperatures, there is a single disordered phase. When there is a stronger interaction between the water and the head group of the amphiphile, lamellar phases appear at large amphiphile concentrations. As this interaction is increased, these phases become more prominent.

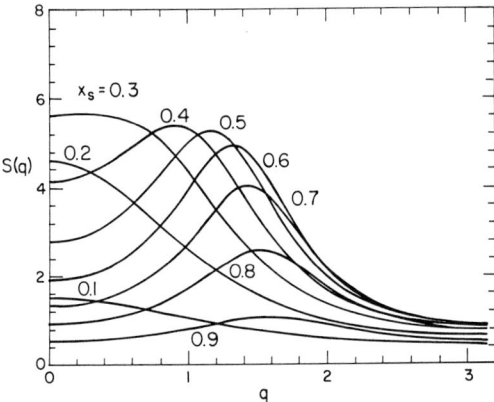

Fig. 2.28 Scattering intensity of a disordered solution of water and amphiphile as obtained from the model (2.21)–(2.24) with $\gamma_1 = \gamma_2 = 0.7$, and $T/\gamma_2 = 4/3$. The scattering intensity is shown for various amphiphile concentrations, x_s, as indicated. From Gompper and Schick (1989b).

The appearance of spatially modulated phases already implies that the disordered phase can be structured. As in the three-component case, this structure can be studied by calculating the scattering intensity of this phase (Gompper and Schick, 1989b). The result is shown in Fig. 2.28. The similarities with the scattering intensity from the microemulsion of the ternary system, as shown in Fig. 2.10 are striking:

(i) At low amphiphile concentration, the scattering peak occurs at wave-vector $q = 0$. With increasing amphiphile concentration, a Lifshitz line is crossed at which the peak moves off of zero wavevector.

(ii) As the amphiphile concentration increases further, the peak moves out, while its intensity decreases.

The above model cannot describe the coexistence region between water- and amphiphile-rich phases which is in the form of a closed loop with a lower critical endpoint. As noted earlier, this is thought to derive from the orientational interactions between water and amphiphile (Goldstein, 1985, and references therein). To include the effects of such interactions, Gompper and Schick (1989b) added temperature-dependent interactions to their model. The influence of orientational interactions between amphiphiles was described not by the $Q_{3,ij}$ and $Q_{4,ij}$ two-particle, vector-interaction terms of the Matsen-Sullivan model (2.26), but by the simpler four-particle, scalar interaction of (2.8), with the (ad hoc) temperature dependence $W = \ln[\sinh(\beta W_0)/\beta W_0]$. Lastly, the reduction of the amphiphile chain entropy, for an amphiphile of length M, was included by means of the

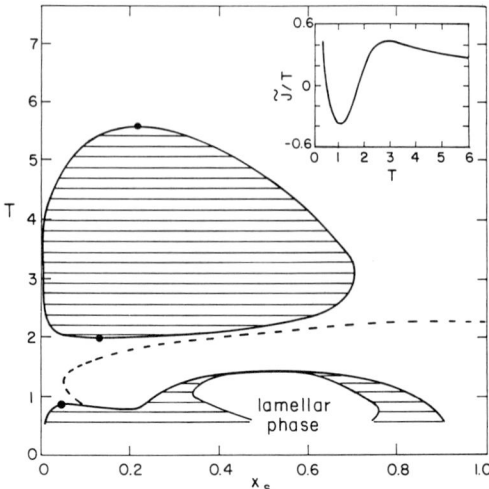

Fig. 2.29 Phase diagram for the model (2.21)–(2.24) including the four-particle term of (2.8) and temperature-dependent effective interactions; $M = 10$, $\gamma_1 = \gamma_2 = 0.7$ and $W_0 = 1.0$. Dots denote critical points, and the dashed line is the locus of points at which the peak in the scattering intensity begins to move from zero wavenumber. The inset shows the temperature dependence of the isotropic coupling constant $\tilde{J}(T)$. From Gompper and Schick (1989b).

Flory–Huggins form of the entropy of mixing (de Gennes, 1979). With these additions, a phase diagram was obtained in which lamellar phases exist at low temperature, a closed coexistence loop of two homogeneous phases occurs at high temperatures, and a disordered phase exists between them. This phase diagram, shown in Fig. 2.29, resembles that of the system $C_{12}E_8$ and water (Tiddy, 1980), for which scattering experiments have been carried out (Degiorgio *et al.*, 1988). A comparison of the peak position and the peak intensity, as obtained from theory and experiment, shows good qualitative agreement (Gompper and Schick, 1989b). Note from this figure that for T a little below 1, the calculation produces a disordered phase, coexisting with the water-rich phase, which is structured and yet consists of 92% water. With an increase in amphiphile concentration, a transition to a lamellar phase occurs. Thus it is tempting to speculate that the structure of this phase is very similar to that of the sponge phase which, in $C_{12}E_5$, exists between 70 and 99.5% water, as was shown in Fig. 1.9.

At low temperatures, experiment shows the existence of hexagonal and cubic as well as lamellar phases as seen in Fig. 1.9. Dawson and Kurtović (1990) added to the Matsen–Sullivan model a term of the form $(\tau_i \cdot \tau_j)^2$, and searched for various lyotropic phases in the zero-temperature limit . With the knowledge of their location at zero temperature, they could extend the

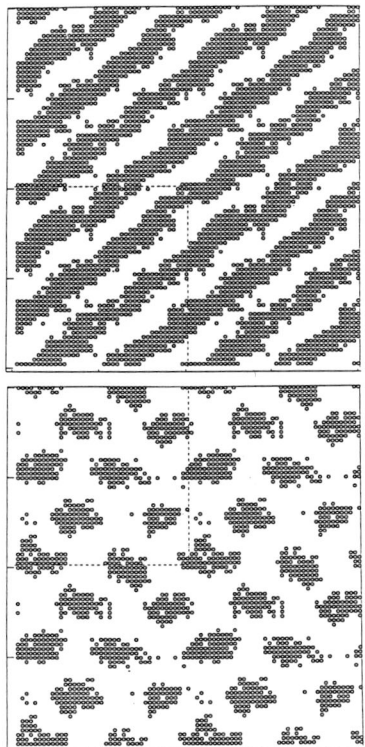

Fig. 2.30 Two mutually perpendicular slices of a $40 \times 40 \times 40$ lattice containing water and the amphiphile H_4T_4. The concentration of the latter is 0.60, and $w = 0.1308$. From Larson (1992).

calculations to finite temperatures using mean-field theory. They do find hexagonal and cubic, as well as lamellar phases. However, with increasing amphiphile concentration, the latter is encountered before the other phases, which is contrary to experiment. Further, minor changes in the choice of their interaction constants produce large shifts in the phase boundaries, making it difficult to determine the systematics of the phase diagram.

Larson (1988, 1992) also obtains hexagonal phases in his model of water and H_4T_4 between concentrations of the latter of 0.45 to 0.65, as shown in the phase diagram of Fig. 2.26. A typical configuration in the hexagonal phase is shown in Fig. 2.30. On the low amphiphile-density side, the hexagonal phase coexists with a disordered phase with spherical micelles. On the high-density side, it coexists with a lamellar phase which appears for amphiphile concentrations beyond 0.73. The lamellae in this phase are perforated by holes, as shown in Fig. 2.31.

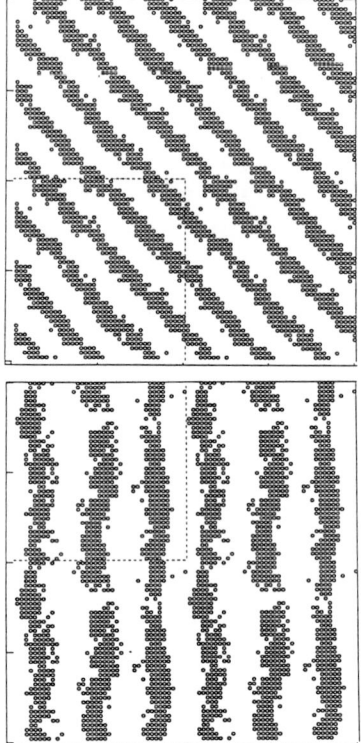

Fig. 2.31 Two mutually perpendicular slices of a $40 \times 40 \times 40$ lattice containing water and the amphiphile $H_4 T_4$. The concentration of the latter is 0.80, and $w = 0.1347$. From Larson (1992).

At low temperatures and amphiphile concentrations between 0.73 and 0.78, the holes in the lamellae order hexagonally. This is but one of many different forms of lamellar phases which can exist, and which are somewhat analogous to different forms of smectic liquid crystals. It is thought that such arrays of ordered holes account for recent experimental observations in a diblock copolymer polymer system (Almdal *et al.*, 1992). It will be an interesting experimental challenge to confirm the existence of these phases in either the systems of polymers or amphiphiles.

2.5 Summary

Having now concluded our review of microscopic lattice models, it is appropriate that we summarize what has been accomplished with them, particularly in light of the experimental behaviour which was described in Section 1.

The phase behaviour obtained from these models is reasonable. By starting with a model containing three components, one is essentially guaranteed the two- to three- to two-phase coexistence which is observed experimentally. As the model is microscopic, one is also able to follow the evolution of the system as the strength of the amphiphile which it contains is changed from weak to strong. Thereby one shows by calculation that systems which are good solubilizers also produce lamellar phases and low interfacial tensions. One also finds that the existence of these lamellar phases causes the disordered phase to be isolated, as in Fig. 2.6, just as in experiment, Fig. 1.7. The isolated region of disordered phase is shown to behave as a microemulsion by an investigation of its structure factors. These show a peak at non-zero wavevector which moves out with increasing amphiphile concentration as in experiment.

As reasonable as are the phase diagrams which have been produced, their relation to experimental ones is, for the most part, indirect. That is, most of them are given as functions of theoretical interaction parameters, and few look like any directly measured phase diagram. However, as we have seen above for the case of systems containing a weak amphiphile, a quite realistic phase diagram, Figs. 2.23 and 2.24, as a function of temperature and amphiphile concentration has been calculated. A portion of the disordered phase could be identified as a microemulsion either according to its structure function, or according to the region in which it was bicontinuous. The various cmcs were calculated and the middle phase followed with temperature. It was observed to emerge as a micellar solution from the lower critical endpoint, rapidly evolve into a bicontinuous microemulsion, and remain so until temperatures very near the upper critical point at which it became an inverse micellar solution, finally disappearing at the upper critical endpoint. There is little reason to doubt that similar work can be carried out for strong amphiphilic systems as well, albeit with much more labour.

Interfacial tensions are readily calculated in microscopic theories, and reductions of at least two orders of magnitude have been obtained. The correlation between amphiphiles which are good solubilizers and those which produce low surface tensions, a correlation expected to hold rather generally, does emerge from these microscopic calculations. The origin of the low tensions in these systems is unrelated to continuous bulk transitions. Rather it can be traced to the fact that a phase of coherent regions of oil and water and very little amphiphile can coexist with water- and oil-rich phases. From the calculated interfacial tensions, the wetting behaviour of the system can be obtained. It is found that the experimental wetting behaviour of the oil/water interface is reproduced by these models with a wetting transition occurring near the disorder line.

There has been less work on the phase diagrams of binary water–amphiphile systems, but reasonable ones have been obtained for the weaker amphiphilic systems. Again a region of the disordered phase has been shown to exhibit a structure function with a peak at a non-zero wavevector. It is reasonable to identify this region with the sponge phase. However, no phase diagram like the one of $C_{12}E_5$, Fig. 1.9, an extreme example exhibiting such a phase, has been produced. Lyotropic phases have been observed in simulations and occur in the sequence encountered experimentally. The detail in which they can be studied is evidenced by the observation of a transition between lamellar phases with ordered and disordered arrays of holes.

3 Ginzburg–Landau theory

3.1 Introduction

As we have seen, microscopic theories produce a wide variety of phase diagrams which result from different choices of the several interaction energies these theories contain. One of the main functions of Ginzburg–Landau theories is to clarify the origins of these different behaviours and to illuminate their universal features. The basic variables of these theories are spatially-varying order-parameter fields which, in contrast to the microscopic degrees of freedom, are defined only on some length scale which is large compared with microscopic lengths, yet small compared with macroscopic ones. In terms of these order parameters, a free energy is constructed from which all thermodynamic information can be obtained. Sometimes, such a Landau free energy can be derived by some procedure in which degrees of freedom on very small length scales are eliminated, while only those on larger length scales are retained. More often, the free energy is simply constructed from symmetry considerations. A good example is provided by the mixing–demixing transition of a binary fluid, such as a mixture of oil and water. One assumes that the essence of the transition can be described in terms of the concentration difference between the two components, and therefore introduces a scalar order parameter $\phi(\mathbf{r})$ which measures this difference locally. The free-energy functional is assumed to have the form

$$\mathcal{F}[\phi(\mathbf{r})] = \int d^3r \,\{g_0(\nabla\phi)^2 + f(\phi) - \mu\phi\}. \tag{3.1}$$

Here, $f(\phi)$ is the free-energy density of a system with a spatially *homogeneous* order parameter, and μ is the chemical potential difference between oil and water. In the vicinity of a critical point where the order parameter is small, the free-energy density f is approximated by the first few terms of a power series in the order parameter ϕ,

$$f(\phi) = a_2\phi^2 + a_3\phi^3 + a_4\phi^4. \tag{3.2}$$

(A linear term is absent as it already appears explicitly in (3.1).) Thermodynamic stability of homogeneous phases requires that $a_4 > 0$. For a system which is symmetric under the interchange of the two components, this symmetry requires that the coefficient a_3 vanish. In this case, the critical point occurs when $a_2 = 0$, $\mu = 0$. When $a_2 > 0$, there is a single, disordered phase (a homogeneous mixture of oil and water), while for $a_2 < 0$ there is coexistence of oil-rich and water-rich phases at $\mu = 0$. Thus the general phase behaviour is seen to depend on only two quantities, μ and the parameter a_2, and is independent of the details of the microscopic interactions. This is a great simplification. By the same token, properties of a specific system, such as the transition temperature and its dependence on these interactions, can only be obtained if a_2 can be derived from a microscopic theory.

From the functional $\mathcal{F}$, thermodynamic quantities and correlation functions can be obtained by functional integration. For example, the two-point correlation function, important for the calculation of the scattering intensity, is given by

$$\langle\phi(\mathbf{r}_1)\phi(\mathbf{r}_2)\rangle = \frac{\int D\{\phi\}\phi(\mathbf{r}_1)\phi(\mathbf{r}_2)\exp\{-\mathcal{F}[\phi(\mathbf{r})]\}}{\int D\{\phi\}\exp\{-\mathcal{F}[\phi(\mathbf{r})]\}}. \tag{3.3}$$

The functional integral can usually not be evaluated exactly, so that various approximation methods are used to deal with it. The simplest and most useful is the mean-field approximation, in which the functional integral is replaced by the maximum of the integrand. This is equivalent to finding the minimum of $\mathcal{F}$, which satisfies the mean-field equation $\delta\mathcal{F}/\delta\phi(\mathbf{r}) = 0$, or

$$2g_0\nabla^2\phi(\mathbf{r}) - f'(\phi(\mathbf{r})) + \mu = 0, \tag{3.4}$$

where $f' = \mathrm{d}f/\mathrm{d}\phi$. For order-parameter fields which vary only in one spatial direction, z, the formal analogy of (3.4) with Newton's equation of motion for a particle of mass $2g_0$ subject to the potential $-f + \mu\phi$ can be used to write the first integral (energy conservation)

$$g_0\left[\phi'(z)\right]^2 - f(\phi(z)) + \mu\phi = \mathrm{const}, \tag{3.5}$$

where $\phi' = \mathrm{d}\phi/\mathrm{d}z$. In fact, it can be shown very generally that only order-parameter fields which vary in one spatial direction can minimize the free-energy functional (3.1) (Derrick, 1964). In the case considered here, $\phi(z)$ is the interfacial profile between the oil-rich and the water-rich phase at coexistence.

For the calculation of correlation functions and scattering intensities, which describe the fluctuations of the order parameter around its average value, one must go beyond the mean-field approximation. The simplest method is to expand the free-energy functional $\mathcal{F}$ to second order in the deviation of the order parameter $\phi(\mathbf{r})$ from its bulk mean-field value $\bar{\phi}(\mathbf{r})$. In terms of $\phi(\mathbf{q})$, the Fourier transform of this deviation, this approximation to the free-energy functional has the form

$$\mathcal{F}[\phi] \approx \mathcal{F}[\bar{\phi}] + \int d^3q K(\mathbf{q})\phi(\mathbf{q})\phi(-\mathbf{q}).$$ (3.6)

With this approximation for the free-energy functional, the scattering intensity can be written

$$S(\mathbf{q}) \equiv \int d^3r_1 \int d^3r_2 \left\{ \langle \phi(\mathbf{r}_1)\phi(\mathbf{r}_2) \rangle - \langle \phi(\mathbf{r}_1) \rangle \langle \phi(\mathbf{r}_2) \rangle \right\} e^{i\mathbf{q}(\mathbf{r}_1 - r_2)}$$

$$= [2K(\mathbf{q})]^{-1}.$$ (3.7)

For a spatially homogeneous phase for which $\bar{\phi}$ is constant, this yields the simple Lorentzian shape

$$S(q) = \frac{1}{f''(\bar{\phi}) + 2g_0 q^2}.$$ (3.8)

3.2 Ginzburg–Landau models of amphiphilic systems

When constructing a Landau free energy to describe oil–water–amphiphile mixtures, one is faced with the same decisions as in formulating a microscopic model: which degrees of freedom are to be retained as essential for the description of the phenomena, and which can be ignored. The simplest model results when one keeps only a single scalar order parameter ϕ, which describes the concentration difference between oil and water.[†] A second choice would be to incorporate as well an additional scalar order parameter to describe the density of amphiphile. These two scalar parameters would permit the description of any combination of densities in the three-component system. If the directional properties of the amphiphile are to be incorporated, then a third, vector order parameter τ can be introduced. Clearly this process can be continued to encompass any degree of complexity desired. We expand on several of these choices below.

[†]For short reviews about a Ginzburg–Landau model with a single scalar order parameter see Gompper (1992) and Gompper and Zschocke (1992a).

3.2.1 One-order-parameter model

The Ginzburg–Landau model (3.1) for oil–water mixtures, with a single scalar order parameter ϕ representing the local concentration difference between oil and water, is inadequate for the description of a ternary mixture which includes a strong amphiphile. One reason that this is so is that the theory predicts the middle phase will always wet the interface between the oil- and water-rich phases (Rowlinson and Widom, 1982), a prediction contrary to the results of experiment (Section 1.2.2). It is easy to see that this prediction emerges from the Landau model by employing the analogy with the motion of a classical particle moving in the potential $-f(\phi) + \mu\phi$, an analogy which emerges from (3.4). The position of the particle at time z is $\phi(z)$. In the ternary mixture problem, we want to consider three-phase coexistence, so that $f(\phi)$ must be characterized by three minima at $\phi_1 < \phi_2 < \phi_3$, corresponding to the three bulk phases, oil-rich, middle and water-rich respectively. Three-phase coexistence occurs when $\mu = 0$ and the depths of the three minima are the same. We choose their common value to be zero, $f(\phi_1) = f(\phi_2) = f(\phi_3) = 0$. With this choice, the constant on the right-hand side of (3.5) is zero. In the classical mechanics analogy, we consider the motion of a particle, with energy zero, moving in a potential characterized by three peaks of equal height. The potential energy at the top of each hill is zero. Calculating the interfacial profile, $\phi(z)$, between the oil-rich phase which extends to $z \to -\infty$, and the water-rich phase which extends to $z \to \infty$, is equivalent to obtaining that trajectory which begins at time $z \to -\infty$ at the top of the hill at ϕ_1 and ends at time $z \to \infty$ at the top of the hill at ϕ_3. Because the particle must pass the top of the hill at ϕ_2, and this hill is of the same height as the others, conservation of energy dictates that the particle spend an infinite amount of time there before continuing on to the third and final hill. In the ternary mixture problem, this corresponds to there being an infinite thickness of middle phase, ϕ_2, between the oil-rich and water-rich phases; i.e. the middle phase wets the oil/water interface. Again, this prediction is a consequence of the Landau free energy of (3.1).

A simple way to avoid this unfortunate feature and yet retain a one-component, scalar order-parameter theory, is suggested by the form of the free energy proposed by Teubner and Strey (1987) to describe the *micro-emulsion* phase,

$$\mathcal{F}_{micro} = \int d^3r \left[c(\nabla^2 \phi)^2 + g_0(\nabla\phi)^2 + a_2\phi^2 \right], \tag{3.9}$$

with $g_0 < 0$ and a_2, $c > 0$. The negative g_0 tends to create interfaces, and the positive c stabilizes the system. By following the same steps as in the last section to calculate the scattering intensity, one finds

$$S(q) = \frac{1}{2[a_2 + g_0 q^2 + cq^4]}. \tag{3.10}$$

For negative g_0, this function has a single maximum at *non-zero* wavevector q. Although (3.9) is clearly an expansion in gradients so that this structure factor should not be applicable at large wavenumbers, it actually describes the measured intensity distributions extremely well (Teubner and Strey, 1987; S.-H. Chen *et al.*, 1990). This is probably due to the fact that $S(q)$ is expected to decay at large q like q^{-4} for any phase characterized by an extensive amount of internal interface (Porod, 1951; Debye *et al.*, 1957; Teubner, 1990). This has been verified for microemulsions (Teubner and Strey, 1987; Vonk *et al.*, 1988). Because the form (3.10) incorporates the correct behaviour for small and large q, it provides a good fit for all q irrespective of its derivation.

The order-parameter order-parameter correlation function $G(r)$ is obtained from (3.10) by Fourier transformation. When g_0 is sufficiently small,

$$g_0^2 < g_{do}^2 \equiv 4ca_2, \tag{3.11}$$

one obtains

$$G(r) = \frac{\xi\lambda}{32\pi^2 cr} e^{-r/\xi} \sin\frac{2\pi r}{\lambda}. \tag{3.12}$$

This correlation function displays two characteristic features of microemulsions, the alternating arrangement of water and oil domains, and the absence of long-range order. The *two* length scales, which appear in (3.12), are given by

$$\xi = \left[\frac{1}{2}\left(\frac{a_2}{c}\right)^{1/2} + \frac{1}{4}\frac{g_0}{c}\right]^{-1/2} \tag{3.13}$$

and

$$\frac{\lambda}{2\pi} = \left[\frac{1}{2}\left(\frac{a_2}{c}\right)^{1/2} - \frac{1}{4}\frac{g_0}{c}\right]^{-1/2}. \tag{3.14}$$

The wavelength λ gives the domain size of coherent oil or water regions, and ξ is the correlation length which characterizes the decay of local order.

When g_0 is larger than g_{do}, the correlation function decays monotonically. The line $g_0 = g_{do}$ which separates these two behaviours of the correlation function is denoted the *disorder* line (DOL). If we think of decreasing g_0, then the disorder line is the line at which oscillatory behaviour first appears in the correlation function. When g_0 decreases to zero, the peak in the structure function (3.10) moves off zero wavevector. The line in parameter space at which this occurs, $g_0 = 0$, defines the Lifshitz line (LL). A peak at non-zero wavevector means that the correlation function is dominated by the oscillatory behaviour. The disordered phase is therefore more structured at the Lifshitz line than at the disorder line. Note that the ratio of the two characteristic lengths, $2\pi\xi/\lambda$ vanishes at the disorder line, because λ diverges there, and takes the value unity at the Lifshitz line, as g_0 vanishes there. As g_0 becomes more negative, this ratio increases and would diverge at the value $g_0 = -g_{do}$ at which the correlation length ξ diverges. But this indicates a continuous transition to a lamellar phase, so that the disordered phase is certainly unstable for $g_0 < -g_{do}$. As the transition from disordered to lamellar phase is expected to be first order (Brazovskii, 1975), the disordered phase will actually become unstable before g_0 becomes this negative.

It is straightforward to generalize the free-energy functional of (3.9) which describes the middle phase to one which describes all phases. First, we must replace the free-energy density $a_2\phi^2$ which appears there by a function $f(\phi)$ which has three minima corresponding to the three homogeneous oil-rich, middle and water-rich phases. Next we must recognize that the coefficients c and g_0, which tell us how difficult it is to make spatial variations in the order parameter, are different in different phases. We know from experiment that the peak in the structure factor in the oil and water phases is at zero wavevector, so that g_0 must be sufficiently positive there, while the peak is at non-zero wavevector in the microemulsion, so that g_0 must be negative there. Thus we should choose g_0 to be a function of $\phi(\mathbf{r})$ itself, $g(\phi)$. With these modifications we arrive at the free-energy density (Gompper and Schick, 1990c)

$$\mathcal{F}[\phi] = \int d^3r \{ c\ [\nabla^2\phi(\mathbf{r})]^2 + g(\phi)[\nabla\phi(\mathbf{r})]^2 + f(\phi) - \mu\phi \}. \tag{3.15}$$

We are now in a position to verify that this free-energy functional does not suffer from the defect inherent in that of (3.1), i.e. it does not necessarily predict that the middle phase wets the oil/water interface. To see this, we observe that the order-parameter fields which extremize the functional of (3.15) satisfy the Euler–Lagrange equation

$$-2c\nabla^2\nabla^2\phi + 2g(\phi)\nabla^2\phi + g'(\phi)(\nabla\phi)^2 - f'(\phi) + \mu = 0. \tag{3.16}$$

where $g' = dg/d\phi$ and $f' = df/d\phi$. Due to the first term being proportional to c, this would correspond to an equation of motion of a classical particle which is no longer Newtonian. Again we seek solutions which vary only in *one* spatial direction, z, so that $\nabla = d/dz$. In this case the first integral of (3.38) is

$$2c[\tfrac{1}{2}(\phi'')^2 - \phi'\phi'''] + g(\phi)[\phi']^2 - f(\phi) + \mu\phi = \text{const}, \tag{3.17}$$

(cf. (3.5)). Again choosing $f(\phi_1) = f(\phi_2) = f(\phi_3) = 0$ at three-phase coexistence which occurs at $\mu = 0$, we see that the constant on the right-hand side of (3.17) vanishes as before. Note however that with this equation of motion, the particle need not come to rest at the top of the second hill even though $f(\phi_2) = 0$ there, because the constant of motion can be satisfied with a non-zero ϕ' due to the presence of the new first term proportional to c. Thus the particle passes the hill in finite time which corresponds to a non-wetting solution in the original statistical mechanics problem.

It is possible to derive the coefficients in the expansion (3.15) from the three-component model of Section 2.1.1. The function $f(\phi)$ is obtained from the Landau expansion of Section 2.2.1 and given by (2.38). The coefficient c and function $g(\phi)$ are obtained by expanding the free energy about the uniform solutions, as in Section 2.2.2. The results for these coefficients are given by Lerczak *et al.* (1992). Of particular interest is the function $g(\phi)$ evaluated in the middle phase of a balanced system, for which $\phi = 0$. It is given by

$$g(0) \sim 1 - (\rho_a/\rho_a^c), \qquad (3.18)$$

where ρ_a is the concentration of amphiphile and $\rho_a^c \equiv J/4|L|$ is the concentration of amphiphile for which $g(0)$ vanishes; that is, it is the concentration of amphiphile needed to bring the system to the Lifshitz line. As the strength of the amphiphilic interaction $|L|$ increases, the amount of amphiphile needed to bring structure to the system decreases.

In general, the functions f and g are unknown, so they are expanded as

$$f(\phi) = \sum_{i=2}^{6} a_i \phi^i \qquad (3.19)$$

and

$$g(\phi) = g(0) + g_2 \phi^2. \qquad (3.20)$$

Here, $a_6 > 0$ and $g_2 > 0$ to insure thermodynamic stability of the model. It is often convenient for numerical and analytic calculations to use a piecewise parabolic form[†] of f,

$$f(\phi) = \begin{cases} \omega_+(\phi - \phi_+)^2 & \text{for } \phi_{0+} < \phi \\ \omega_0 \phi^2 + f_0 & \text{for } \phi_{0-} < \phi < \phi_{0+} \\ \omega_-(\phi - \phi_-)^2 & \text{for } \phi < \phi_{0-}, \end{cases} \qquad (3.21a)$$

where ϕ_{0+} and ϕ_{0-} are defined such that f is continuous, and a piecewise constant form of g,

---

[†]Such a piecewise parabolic form of the free-energy density was introduced by Lipowsky (1984) in the context of wetting transitions.

$$g(\phi) = \begin{cases} g_+ & \text{for } \phi_{0+} < \phi \\ g_0 & \text{for } \phi_{0-} < \phi < \phi_{0+} \\ g_- & \text{for } \phi < \phi_{0-}. \end{cases} \quad (3.21b)$$

3.2.2 Two-order-parameter models

Just as the Hamiltonian $\mathcal{H}_0$ of (2.1) for a simple ternary mixture is the natural starting point for a microscopic model with two independent densities, so the Landau expansion it generates is the simplest starting point for a Landau theory with two scalar order parameters. In the case of oil–water symmetry, this functional reads

$$\mathcal{F}_0[\phi, \psi] = \int d^3r [\beta_2 (\nabla^2 \phi)^2 + \beta_1 (\nabla \phi)^2 + \tilde{A}_2 \phi^2 + \tilde{A}_4 \phi^4 + \tilde{A}_6 \phi^6$$
$$+ \delta_1 (\nabla \psi)^2 + \tilde{B}_2 \psi^2 + \tilde{B}_3 \psi^3 + \tilde{B}_4 \psi^4 - \mu_s \psi \quad (3.22)$$
$$+ \tilde{C}_3 \phi^2 \psi + \tilde{C}_4 \phi^2 \psi^2 + ...]$$

for two scalar order-parameter fields, $\phi(\mathbf{r})$ and $\psi(\mathbf{r})$. (Note that this is simply the extension to the case of spatially varying fields of the functional given by (2.37) for uniform fields.) The two fields can be taken to be the local concentration difference of oil and water, and the local concentration of amphiphile. The coefficients of all gradient terms, β_1, β_2, δ_1, are positive. Thermodynamic stability also requires $\tilde{A}_6 > 0$, $\tilde{B}_4 > 0$ and $\tilde{C}_4 > 0$. Just as the properties of the amphiphile have to be added to $\mathcal{H}_0$, so an additional term must be added to $\mathcal{F}_0$ to mimic amphiphilic properties. This can be done by including a term

$$\mathcal{F}_{amp}[\phi, \psi] = \int d^3r \, [\gamma_1 \phi^2 \nabla^2 \psi + \gamma_2 \psi (\nabla \phi)^2] \,, \quad (3.23)$$

which, with $\gamma_1, \gamma_2 < 0$, favours the amphiphile to sit at oil/water interfaces, just as the corresponding term in (2.6) does in the lattice model. Thus we arrive at

$$\mathcal{F}[\phi, \psi] = \mathcal{F}_0[\phi, \psi] + \mathcal{F}_{amp}[\phi, \psi] \,. \quad (3.24)$$

This model is of interest when information about the amphiphile density or the amphiphile scattering intensity is required. It has been used in the special case in which β_2, δ_1, $\tilde{B}_3$, $\tilde{B}_4$, γ_1, and $\tilde{C}_3$ all vanish to study the dynamics of phase separation in the presence of amphiphiles (Laradji et al., 1991b, 1992). Unfortunately, with $\gamma_2 < 0$ and $\beta_2 = 0$ the model is not well defined because the free energy of configurations with high amphiphile concentration ψ and small wavelength oscillations of ϕ is not bounded from below. A similar model has been studied by Anisimov et al. (1992a).

A novel application of the Landau–Ginzburg theory with two scalar order parameters is to the sponge phase of the binary system of amphiphile and water by Roux *et al.* (1990, 1992a). The Landau theory encapsulates ideas expressed previously in terms of lattice models by Cates *et al.* (1988b). The order parameter ψ again is related to the amphiphilic concentration. The order parameter ϕ, which in the ternary system described the concentration difference between the oil and water regions on either side of the amphiphilic monolayer, here describes the concentration difference between water regions on either side of the amphiphilic bilayer. In other words, the use of the order parameter ϕ is a way of incorporating into the description of the fluid the fact that there are bilayers within it. When the average value of ϕ vanishes, the sponge phase is said to be symmetric; when it is non-zero, it is asymmetric. The various predictions of the theory are discussed below in Section 3.4.1.

Another model with two-order parameters is defined by the free-energy functional (K. Chen *et al.*, 1990)

$$\mathcal{F}[\phi,] = \int d^3r \{ \beta_2 (\nabla^2 \phi)^2 + \beta_1 (\nabla \phi)^2 + \tilde{A}_2 \phi^2 + \tilde{A}_4 \phi^4 + \tilde{A}_6 \phi^6 \}$$

$$+ \int d^3r \{ \tilde{B}_2 \tau^2 + \delta_1 (\nabla \cdot \tau)^2$$

$$+ \tilde{C}_4 \phi^2 \tau^2 + \gamma \tau \cdot \nabla \phi \},$$

$$= \mathcal{F}_{bin}[\phi] + \mathcal{F}_{amp}[\tau] + \mathcal{F}_{int}[\phi, \tau]. \tag{3.25}$$

Here, $\phi(\mathbf{r})$ is a scalar order-parameter field, as before, and $\tau(\mathbf{r})$ a *vector* order-parameter field. The first line in (3.25) describes the oil–water subsystem (for oil–water symmetry), with a free-energy functional $\mathcal{F}_{bin}$ where $\beta_1 > 0, \beta_2 > 0$, and $\tilde{A}_4 > 0, \tilde{A}_6 > 0$. The coefficient $\tilde{A}_2$ is chosen to be negative so that oil and water do not mix in the absence of amphiphile. The second line describes the pure amphiphile, with a free-energy functional $\mathcal{F}_{amp}$ where $\tilde{B}_2, \delta_1 > 0$. Only terms quadratic in τ are considered because the ordering of the amphiphiles in the absence of oil or water is not within the scope of this model. Finally, the terms in the last line describe the interactions of the amphiphiles with oil and water by means of a free-energy functional $\mathcal{F}_{int}$. Here, $\gamma \neq 0$ favours amphiphiles to sit preferentially at oil/water interfaces, where $\nabla \phi$ is large. The coefficient $\tilde{C}_4 > 0$ is a measure of the miscibility of amphiphile with pure oil and water.

For $\tilde{C}_4 = 0$, the amphiphilic degrees of freedom can be integrated out exactly. This is most easily done in Fourier space. In this case, one ends up with a single order-parameter model

$$\mathcal{F}[\phi] = \int \frac{d^3q}{(2\pi)^3} \left[\beta_1 q^2 + \beta_2 q^4 - \frac{\gamma^2 q^2}{4(\tilde{B}_2 + \delta_1 q^2)} \right] \phi(\mathbf{q})\phi(-\mathbf{q})$$
$$+ \int d^3r \left[\tilde{A}_2 \phi^2 + \tilde{A}_4 \phi^4 + \tilde{A}_6 \phi^6 \right].$$

(3.26)

The effective interaction can be expanded for small q (which is after all the reason only the first few gradient terms were included in (3.25)), to obtain a functional of the same form as $\mathcal{F}_{bin}$, with the renormalized coefficients

$$\beta_1 \longrightarrow \beta_1 - \frac{\gamma^2}{4\tilde{B}_2}$$

(3.27)

$$\beta_2 \longrightarrow \beta_2 + \frac{\gamma^2 \delta_1}{4\tilde{B}_2^2}.$$

The effect of the amphiphile is therefore to reduce the coefficient of the $(\nabla\phi)^2$ term, exactly as was found in the microscopic derivation of (3.18). For non-zero $\tilde{C}_4$, a one-loop calculation shows (K. Chen *et al.*, 1990) that the coefficients $\tilde{A}_2$ and $\tilde{A}_4$ are renormalized in such a way that the structure of the free-energy density $f(\phi) = \tilde{A}_2 \phi^2 + \tilde{A}_4 \phi^4 + \tilde{A}_6 \phi^6$, which has only two minima when $\tilde{A}_2 < 0$ and $\tilde{A}_4 > 0$, can be changed into one with three minima, $\tilde{A}_2 > 0$, $\tilde{A}_4 < 0$, corresponding to the coexistence between three homogeneous phases. At the same time, a $\phi^2(\nabla\phi)^2$ term is generated (again to lowest order in an expansion in q). Therefore, one arrives at a one scalar order-parameter model which is of exactly the same form discussed in the last section. With the renormalization of (3.27), the coefficients in the expansions of (3.19) and (3.20) can now be interpreted in terms of the interactions of amphiphile heads and tails with oil and water. Furthermore, in the present model, correlation functions involving the densities of amphiphile can be calculated. For example, for $\tilde{C}_4 = 0$ one has

$$\langle \tau(\mathbf{q})\tau(-\mathbf{q}) \rangle = \frac{1}{2(\tilde{B}_2 + \delta_1 q^2)} + \frac{\gamma^2 q^2}{4(\tilde{B}_2 + \delta_1 q^2)^2} \langle \phi(\mathbf{q})\phi(-\mathbf{q}) \rangle.$$

(3.28)

3.2.3 Three-order-parameter model

Finally, models can be constructed in which both the the density and the orientational degrees of the amphiphile are taken into account, so that three-order parameters are employed (Kawasaki and Kawakatsu, 1990; Anisimov *et al.*, 1992b; Gompper and Klein, 1992). The particular version we consider here can be written

$$\mathcal{F}[\phi, \psi, \tau] = \mathcal{F}_0[\phi, \psi] + \mathcal{F}_{amp}[\tau] + \mathcal{F}_{int}[\phi, \psi, \tau] \qquad (3.29)$$

where $\mathcal{F}_0[\phi, \psi]$ is again the free-energy functional of the simple ternary mixture. The functional $\mathcal{F}_{amp}[\tau]$ describes the pure amphiphile subsystem (Gompper and Klein, 1992),

$$\mathcal{F}_{amp}[\tau] = \int d^3r[\alpha_1 (\nabla \cdot \tau)^2 + \alpha_2 (\nabla^2 \tau)^2$$
$$+ \alpha_3 (\nabla \times \tau)^2 + \alpha_4 (\nabla(\nabla \cdot \tau))^2 + \tilde{B}_2 \tau^2 + \tilde{B}_4 \tau^4]. \qquad (3.30)$$

Here, all terms with first and second derivatives are included which are rotationally invariant and quadratic in τ. We require $\alpha_2 > 0$, $\alpha_4 > 0$ and $\tilde{B}_4 > 0$ to ensure thermodynamic stability. Further, $\tilde{B}_2 > 0$ in order to have no homogeneous ordered phases in which all amphiphiles would point in the same direction, a state highly unfavourable due to close contact of amphiphile heads and tails. The second derivatives have been included in the free-energy functional (3.30) in order that the system remain thermodynamically stable irrespective of the sign of the coefficients of the $(\nabla \cdot \tau)^2$ and the $(\nabla \times \tau)^2$.

To inform the choice of α_1 and α_3, the structure functions of the disordered phase are calculated within the Ornstein–Zernike approximation. One finds (Gompper and Klein, 1992)

$$\langle \tau_z(\mathbf{q})\tau_z(-\mathbf{q})\rangle = \frac{1}{\zeta + \epsilon}\left(1 + \frac{\epsilon}{\zeta}\sin^2\theta\right) \qquad (3.31)$$

where

$$\zeta = \tilde{B}_2 + \alpha_3 q^2 + \alpha_2 q^4$$
$$\epsilon = (\alpha_1 - \alpha_3)q^2 + \alpha_4 q^4 \qquad (3.32)$$

and θ is the angle between $\mathbf{q}$ and the z-axis, i.e. $\cos^2\theta = q_z^2/q^2$. The Fourier transform of this function is the probability of finding two amphiphiles oriented in the z direction separated by $\mathbf{r}$. When $\theta = \pi/2$, $\langle \tau_z(\mathbf{q})\tau_z(-\mathbf{q})\rangle = 1/\zeta$ which has a peak at $q = 0$ as long as $\alpha_3 > 0$. Thus one should choose a positive α_3 so that the amphiphiles behave ferromagnetically within a monolayer. When $\theta = 0$, the structure function has a peak at non-zero wavevector for $\alpha_1 < 0$, so that one should choose such a value to describe bilayer formation in which amphiphiles are ordered antiferromagnetically in neighbouring layers.

Finally, $\mathcal{F}_{int}[\phi, \psi, \tau]$ describes the interactions between the local direction of the amphiphile, τ, with the other two order parameters,

$$\mathcal{F}_{int}[\phi, \psi, \tau] = \int d^3r[\gamma\tau \cdot \nabla\phi + \tilde{D}_3\psi\tau^2 + \tilde{D}_4\psi^2\tau^2]. \qquad (3.33)$$

The first term acts to align the amphiphile properly with respect to oil and water, as in the model of K. Chen *et al.* (1990). The second term acts like a local chemical potential for the amphiphile concentration. With $\tilde{D}_3 < 0$ the local concentration of amphiphiles is large when they are oriented. Thermodynamic stability requires $\tilde{D}_4 > 0$.

3.3 Some results of the one-order-parameter model

We shall describe some results which have been obtained from the Ginzburg–Landau model which employs a single scalar order parameter. The free-energy functional, given previously in (3.15), is repeated here for convenience:

$$\mathcal{F}[\phi] = \int d^3r \left[c \, (\nabla^2 \phi(\mathbf{r}))^2 + g(\phi(\mathbf{r})) \, (\nabla \phi(\mathbf{r}))^2 + f(\phi(\mathbf{r})) - \mu \phi(\mathbf{r}) \right]. \quad (3.34)$$

It will be seen that this model is capable of explaining and predicting the basic properties of oil–water–amphiphile mixtures, provided that the concentrations of oil and water are comparable and are large compared with the amphiphile concentration.

3.3.1 The oil/microemulsion and oil/water interfaces at three-phase coexistence

Order-parameter profiles for the oil/water, the oil/microemulsion and the microemulsion/water interfaces can be calculated from the Euler–Lagrange equation (3.16) restricted to variation in one direction only,

$$-2c\phi'''' + 2g(\phi)\phi'' + g'(\phi)[\phi']^2 - f'(\phi) + \mu = 0, \quad (3.35)$$

and its first integral

$$2c \left[\frac{1}{2}(\phi'')^2 - \phi'\phi''' \right] + g(\phi)[\phi']^2 - f(\phi) + \mu\phi = \text{const}. \quad (3.36)$$

The interfacial tensions follow on substitution of these profiles into the free-energy functional (3.34) and subtraction of the bulk free energy. One of the few profiles which can be calculated analytically results from the use of the piecewise parabolic model (3.21). It is the profile between oil (located at $z > 0$) and microemulsion ($z < 0$), and is given by

$$\phi(z) = \begin{cases} \phi_+ - A_1 e^{-\alpha_1 z} - A_2 e^{-\alpha_2 z} & \text{for } z > 0 \\ B_1 e^{+\beta_1 z} + B_2 e^{+\beta_2 z} & \text{for } z < 0, \end{cases} \quad (3.37)$$

where $\alpha_{1,2}$ and $\beta_{1,2}$ are the roots with positive real parts of the equations

$$c\alpha^4 - g_+\alpha^2 + \omega_+ = 0,$$
$$c\beta^4 - g_0\beta^2 + \omega_0 = 0. \tag{3.38}$$

The four amplitudes are determined by the continuity of ϕ and its first two derivatives at $z = 0$ and the condition that $\phi(0) = \phi_{0+}$, which follows from (3.21a). Note that with the profile above, the value of the invariant for $z > 0$ is easily determined by considering the limit $z \to \infty$, by which one finds that it has the value zero. Similarly an evaluation in the limit $z \to -\infty$ shows that it is also zero for $z < 0$. Thus, with the profile above, we need not ensure that the values of the invariant match at $z = 0$. They already do so. The amplitudes are

$$A_1 = +\frac{\phi_{0+}}{Q}\frac{1}{\alpha_1 - \alpha_2}\left[-\sqrt{\frac{\omega_0}{\omega_+}}\alpha_2 Q + \sqrt{\frac{\omega_0}{\omega_+}}\alpha_1\alpha_2 + \beta_1\beta_2\right]$$

$$A_2 = -\frac{\phi_{0+}}{Q}\frac{1}{\alpha_1 - \alpha_2}\left[-\sqrt{\frac{\omega_0}{\omega_+}}\alpha_1 Q + \sqrt{\frac{\omega_0}{\omega_+}}\alpha_1\alpha_2 + \beta_1\beta_2\right]$$

$$B_1 = -\frac{\phi_{0+}}{Q}\frac{1}{\beta_1 - \beta_2}\left[\beta_2 Q - \sqrt{\frac{\omega_0}{\omega_+}}\alpha_1\alpha_2 - \beta_1\beta_2\right]$$

$$B_2 = +\frac{\phi_{0+}}{Q}\frac{1}{\beta_1 - \beta_2}\left[\beta_1 Q - \sqrt{\frac{\omega_0}{\omega_+}}\alpha_1\alpha_2 - \beta_1\beta_2\right]$$

$$\tag{3.39}$$

where

$$Q = \alpha_1 + \alpha_2 + \beta_1 + \beta_2. \tag{3.40}$$

For all values of $g_0 < g_{do}$ (and $g_+ > g_{do}$), the order-parameter profile reaches its bulk value on the oil side monotonically, but oscillates on the microemulsion side resulting in a water-rich layer at the oil/microemulsion interface. This can be seen clearly in Fig. 3.1(a).

Because the bulk free energy of the coexisting phases has been taken to be zero in (3.21a), the interfacial tension σ_{om} of the oil/microemulsion interface is simply the value of the free-energy functional (3.34) obtained on insertion of the order-parameter profile:

$$\sigma_{om} = g_+(\alpha_1 A_1^2 + \alpha_2 A_2^2) + \left(2\omega_+ + g_+\sqrt{\frac{\omega_+}{c}}\right)\frac{2A_1 A_2}{\alpha_1 + \alpha_2}$$

$$+ g_0(\beta_1 B_1^2 + \beta_2 B_2^2) + \left(2\omega_0 + g_0\sqrt{\frac{\omega_0}{c}}\right)\frac{2B_1 B_2}{\beta_1 + \beta_2}.$$

$$\tag{3.41}$$

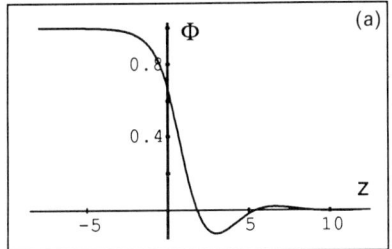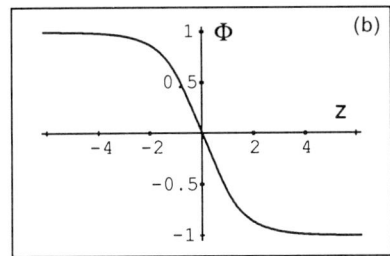

Fig. 3.1 Order-parameter profile of (a) the oil/microemulsion and (b) the oil/water interfaces at three-phase coexistence for $g_0 < g_{do}$. The profiles are calculated from model (3.34), (3.21) with the parameters $c = 1, g_+ = g_- = 4.5, g_0 = -1, \omega_+ = \omega_- = 4,$ $\omega_0 = 1, \phi_+ = 1, \phi_- = -1, f_0 = 0, \mu = 0$. From Gompper (1992). Reprinted by permission of Kluwer Academic Publishers.

At the disorder line, the exponents β_1 and β_2 and the amplitudes B_1 and B_2 are singular. The correlation length of the order-parameter correlation function is also singular. However, as the disorder line is approached the second largest correlation length becomes comparable to the largest, and equals it at the disorder line itself. This second length has the same singularity as the correlation length, and these singularities cancel in the calculation of the surface and bulk free energies. Thus these free energies are analytic at the disorder line.

Whereas the oil/microemulsion and water/microemulsion profiles can be calculated analytically using the piecewise parabolic model of (3.21), the oil/water profile cannot. In the case of oil–water symmetry, this profile has the general form

$$\phi(z) = \begin{cases} \phi_+ - A_1' e^{-\alpha_1 z} - A_2' e^{-\alpha_2 z} & \text{for } z > \ell/2 \\ B_1' \sinh(\beta_1 z) + B_2' \sinh(\beta_2 z) & \text{for } -\ell/2 < z < \ell/2 \\ -\phi_+ + A_1' e^{+\alpha_1 z} + A_2' e^{+\alpha_2 z} & \text{for } z < -\ell/2 \end{cases} \quad (3.42)$$

To determine the four amplitudes and the unknown thickness we have the conditions of continuity of ϕ and its first two derivatives at $\ell/2$, and the condition that $\phi(\ell/2) = \phi_{0+}$. A fifth condition arises because the value of the invariant in the region $z < |\ell/2|$ is not determined by the form of the profile, and must be set to its zero value outside this region, the value following from the form of the profile there. The four amplitudes can be calculated analytically in terms of the unknown ℓ, but this last must be determined numerically. For $g_0 < g_{do}$ the resulting profile is shown in Fig. 3.1(b). From this profile, we see that the oil/water interface is not wetted by the microemulsion in this case. A similar calculation for $g_0 > g_{do}$ gives a profile which clearly shows that the oil/water interface is wetted by the microemulsion.

In the first case, the disordered fluid has an oscillating correlation function, and is therefore a microemulsion by our definition. Hence the microemulsion, so defined, does not wet the oil/water interface. Since we have seen that a fluid with a scattering peak at non-zero q always has an oscillating correlation function (but not *vice versa*), it follows that a fluid which wets the oil/water interface should have a peak at $q = 0$, and that a fluid which has a peak in $S(q)$ at non-zero q should not wet the oil/water interface. This conclusion is modified by the presence of long-range van der Waals forces, as shown below.

3.3.2   Wetting transitions at the oil/water interface

In the theory of wetting transitions the concept of the effective interface potential has turned out to be very useful (Dietrich, 1988; Schick, 1990; Forgacs *et al.*, 1991, and references therein). For the present problem, the effective interaction between an oil/microemulsion and a microemulsion/water interface at distance ℓ is to be calculated. If the global minimum of the interaction potential $V(\ell)$ occurs for $\ell \to \infty$, corresponding to a macroscopically thick layer of middle phase between the oil- and water-rich phases, then the oil/water interface is wetted by the middle phase. If the global minimum occurs for a finite value of ℓ, then the interface is not wetted by the middle phase. The effective interface potential is easily obtained because, as noted earlier, analytic expressions for the amplitudes of the profile (3.42) as functions of ℓ can be calculated. Instead of solving for the thickness numerically, we leave it as a free parameter. Substitution of the profile into the functional (3.34) produces

$$\sigma(\ell) = \sigma_{om} + \sigma_{mw} + V(\ell), \tag{3.43}$$

with the first two terms on the right being the interfacial tensions between oil and middle phases, and between middle and water phases. The equilibrium value of the oil/water tension is $\sigma_{ow} = \min \sigma(\ell)$, and the minimum value of the effective interface potential must be negative or zero from the stability argument of (1.1). The effective interaction is shown in Fig. 3.2.

The asymptotic form of the interaction potential at large separations is found to be (Gompper and Schick, 1990c)

$$V(\ell) \to C_1 e^{-\ell/\xi} \cos(k\ell + \theta), \tag{3.44}$$

which is also shown in Fig. 3.2. Here θ is an unimportant phase, $\xi^{-1} = \min(\mathrm{Re}(\beta_1), \mathrm{Re}(\beta_2))$, and $k = \mathrm{Im}(\beta_1) = \mathrm{Im}(\beta_2)$ are the correlation length and the characteristic wavevector of the *bulk* correlation function, respectively. Thus the oscillations of the effective interface potential reflect the oscillations of the bulk correlation function. An intuitive explanation for

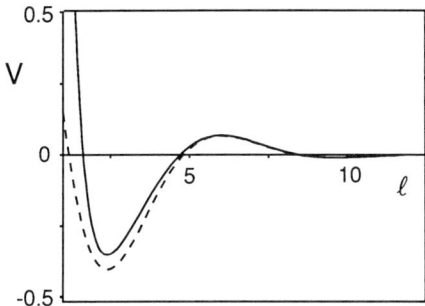

Fig. 3.2 Effective interaction potential between an oil/microemulsion and a micro-emulsion/water interface separated by a distance ℓ. The parameters of the model (3.21) are the same as in Fig. 3.1. Dashed line is the approximation of (3.44).

this behaviour is apparent when one considers the oscillatory profiles between middle phase and oil or middle phase and water shown in Fig. 3.1. If one considers putting two such profiles together, it is not difficult to see that the energy will oscillate as oil- and water-rich regions do, or do not, coincide. As long as $V(\ell)$ has the form (3.44) with $k > 0$, the potential has an absolute minimum at finite ℓ, and the oil/water interface is not wet. As the disorder line is approached, the wavevector $k \to 0$, and the minimum of $V(\ell)$ moves to larger and larger values of ℓ, diverging at the disorder line as $\langle \ell \rangle \sim \delta^{-1/2}$, where $\delta = g_{do} - g_0$. This form, arising from the analyticity of the bulk-free energy at the disorder line, should be compared with the logarithmic divergence usually seen in systems with short-range interactions. From (3.44), the oil/water interfacial tension exhibits an essential singularity $\exp(-\text{const}.\delta^{-1/2})$.

In physical systems, long-range van der Waals forces are always present. These forces, which couple to the *average* density of the fluid, produce an additional term in the interface potential of (3.44), one which varies as $W\ell^{-p}$, with $p = 2$ (Dietrich, 1988). Here, W is the Hamaker constant; it is propor-tional to the electron density difference of the phases involved,

$$W \sim \frac{(\rho_o - \rho_m)(\rho_m - \rho_w)}{(\rho_o - \rho_w)}. \tag{3.45}$$

In the case considered here, the electron density of the middle phase, ρ_m, lies between the electron densities of the other two phases, $\rho_w > \rho_m > \rho_o$, so that $W > 0$, and the term $W\ell^{-2}$ is positive. The effect of the new term is shown schematically in Fig. 3.3. It causes the wetting transition to become *first order*. This transition does not take place at the disorder line, but on the micro-emulsion side of it. Its exact location depends on the value of the Hamaker constant for the system under consideration. Because the transition now

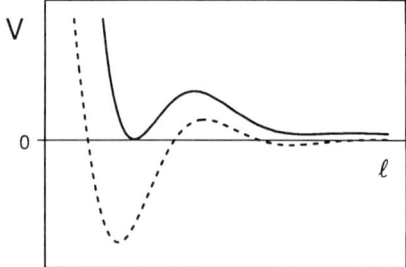

Fig. 3.3 Effective interaction potential between an oil/microemulsion and a micro-emulsion/water interface separated by a distance ℓ, with (full line) and without (dashed line) the contribution from the van der Waals interactions (schematic).

occurs on the microemulsion side of the disorder line where there may or may not be a peak at non-zero wavevector in the structure function, one can make no simple correlation between the location of this peak and the wetting or non-wetting of the interface.

Several experiments largely confirm the theoretical expectations. First, non-monotonic interfacial profiles[†] have been reported by Meunier (1985). Second, there is little doubt that balanced microemulsions with long-chain amphiphiles do not wet the oil/water interface (Widom, 1987; Kahlweit et al., 1987). Third, it is also well established that as the tricritical point is approached, for example by mixing short- and long-chain amphiphiles and increasing the concentration of the former, the disorder line is crossed and a wetting transition occurs (Kahlweit et al., 1991). The correlation between the scattering behaviour and the wetting properties have been studied by two groups. Schubert and Strey (1991) examined three different amphiphiles, C_8E_3, C_6E_2 and C_4E_1, in octane and water mixtures. They weakened the strength of the amphiphiles by adding formamide and could thereby induce a wetting transition at the oil/water interface. All samples were at equal volume fractions of polar and non-polar components. They fitted the structure function from small-angle neutron scattering on these systems to the Teubner–Strey form of (3.10), and thereby extracted the coefficients a_2, c and g_0. From the first two, they obtained g_{do} of (3.11), the value of g_0 at the disorder line. By comparing this to the measured value of g_0, they knew where the system was with respect to the disorder line. The wetting transitions they found occurred near the Lifshitz line, well on the microemulsion side of the disorder line, in agreement with predictions. As the wetting transition is approached from the non-wetting side, the peak in the structure function

---

[†]Non-monotonic profiles have been measured at the interface between air and mixtures of water and amphiphile by Pershan (1989), Cevc et al. (1990), and at the interface between air and microemulsions by Zhou et al. (1992).

ultimately moves toward zero wavevector because the system is becoming less strongly ordered. Abillon et al. (1991) studied the system of C_6E_2 in hexadecane and water and induced a wetting transition by varying the temperature. They observed the wetting transition to occur while there was still an observable peak in the structure function. Hence it occurs on the microemulsion side of the Lifshitz line and, therefore, on that side of the disorder line as predicted. They did not observe the peak in the structure factor to move towards zero as the wetting transition was approached, as did Schubert and Strey (1991), but they approached the wetting transition along a different experimental path, one in which the temperature, and hence the composition of the middle phase, was varied. As noted above, when long-range forces are included, there are few general statements that one can make relating the structure function to the wetting transition.

The majority of wetting transitions which have been observed in ternary amphiphilic mixtures have been brought about by varying the temperature, so that one is approaching a critical endpoint (Robert and Jeng, 1988; Smith and Covatch, 1990; Aratono and Kahlweit, 1991, 1992). According to a famous argument of Cahn (1977), one expects a wetting transition to occur before the endpoint is reached. One might expect that the contact angle would decrease uniformly as the endpoint was approached. The experimental results are contrary to these expectations, however (Aratono and Kahlweit, 1991, 1992). While the decrease is monotonic for weaker amphiphiles, it is decidedly non-monotonic for stronger ones, as shown in Fig. 3.4(a).

In an attempt to understand this, Putz et al. (1992, 1993) again employed the piecewise parabolic model to study the wetting behaviour as the critical endpoint is approached. Their result is shown in Fig. 3.4(b). The contact angle of strong amphiphiles shows the non-monotonic behaviour observed in experiment. As the critical point is approached, the middle phase always wets, in agreement with the Cahn argument, although the wetting occurs closer and closer to the critical endpoint for stronger and stronger amphiphiles.

The reason for the non-monotonic behaviour of the contact angle is not difficult to discern. Away from the endpoints, the middle phase is strongly structured so that the value of $g(\phi)$ in this phase, $g(\phi_{mid}) \equiv g(0)$, is negative. Thus large gradients of the density are favoured by the layer of amphiphile at the interface. As one takes the system towards either of its critical endpoints, the value of $g(\phi)$ at the disorder line (3.11), $g_{do} = \sqrt{4ca_2}$, decreases because a_2 vanishes at the endpoint. Hence the negative value of $g(0)$, which in the microemulsion is restricted to the interval $-g_{do} < g(0) \leq g_{do}$, is closer to its lower stability limit which implies that the microemulsion becomes more structured. Consequently, the contact angle increases. If it increased to π, the water phase would wet the interface between oil and middle phases, replacing one gradient with two larger ones. This process cannot continue

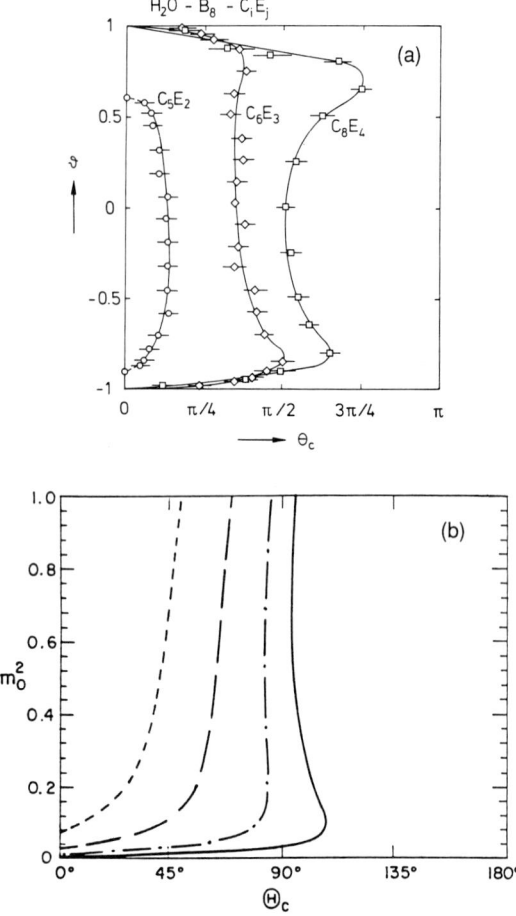

Fig. 3.4 (a) Contact angle Θ_c vs. reduced temperature ϑ for systems of water, n-octane, and three different amphiphiles. The critical endpoints are at $\vartheta = \pm 1$. From Aratono and Kahlweit (1992). (b) Calculated behaviour of the contact angle vs. the square of the reduced order parameter m_0^2 which, in mean-field theory, is proportional to the reduced temperature. Results for a sequence of four amphiphiles, with strengths increasing from left to right, are shown. From Putz *et al.* (1992).

all the way to the endpoints, however, for if $g(0)$ remained negative in the middle phase, eventually one would have $g(0) = -g_{do}$ and, as we have argued above, this would imply that a transition to a lamellar phase would be encountered. As this is not the case experimentally, it follows that $g(0)$ must become positive, so that sufficiently close to the endpoint, $g(0) > g_{do}$. Hence the fluids which are becoming critical there are unstructured. In this case, gradients in the density are unfavourable. The contact angle must decrease and, as shown earlier, a wetting transition occurs as $g(0)$ approaches g_{do} from below at which point the system replaces one large change of density, from the oil to the water phase, by two smaller ones which cost less free energy.

From the experimental results in Fig. 3.4(a), one can see that the transition is first order. This follows from the fact that on approaching a first-order wetting transition, the contact angle should vanish with infinite slope and positive curvature when plotted against the temperature, whereas on approaching a continuous transition, it vanishes with a zero slope and negative curvature (Dietrich, 1988). Clearly the former scenario fits the data. That the wetting transition is first order was also reported by Robert and Jeng (1988). This is in agreement with the prediction of the Landau theory.

One also sees from Fig. 3.4(a) that as the amphiphile gets stronger, the wetting transition occurs closer and closer to the endpoint. The question naturally arises as to whether the transition might not occur at all as the endpoint was approached. As we have argued that a structured middle phase for which $-g_{do} < g(0) \leq g_{do}$ will not wet the interface, and that g_{do} vanishes at the endpoint, it is clear that we must also have $g(0)$ vanish at the endpoint if the wetting transition is not to occur. If this were the case, we would have a Lifshitz critical endpoint (Hornreich et al., 1975). At such a point, two of the fluid phases and a lamellar phase become critical simultaneously. To investigate this, Gompper et al. (1991) employed the one-order-parameter model with

$$f(\phi) = \omega\phi^2(\phi - \phi_+)^2(\phi - \phi_-)^2, \tag{3.46}$$

$$g(\phi) = g(0) + g_2\phi^2. \tag{3.47}$$

Critical endpoints occur when $\phi_- \to 0$ while $\phi_+ \neq 0$ and *vice versa*. The endpoint is a Lifshitz one if $g(0) = 0$. All interfacial tensions were evaluated by solving (3.17) (with $\mu = 0$) numerically, and evaluating the free energy (3.15) with the profiles. The results showed that the interface was not wet away from the Lifshitz critical endpoint, but it could not be approached sufficiently closely to determine that the non-wetting remained to the transition itself.

More successful was the investigation of the approach to a Lifshitz tricritical point (Aharony et al., 1987) using the same f and g above. A tricritical point is described if ϕ_+ and ϕ_- vanish simultaneously. Due to the vanishing of g_{do} at the tricritical point, one has a scenario similar to that at a critical endpoint. If $g(0) > 0$, the condition $g(0) > g_{do}$ must be reached as the tricritical point is approached. The middle phase becomes structureless and the tricritical behaviour is ordinary. It has been shown that a

usual square-gradient Landau model (with $c = 0$, $g(\phi) = $ const, and $f(\phi)$
given by (3.46)) describes the experimental data near the tricritical point
very well (Kleinert, 1986a). This implies in particular that the microemulsion
must wet the oil/water interface as the tricritical point is approached (Widom,
1987; Gompper et al., 1991), a behaviour clearly demonstrated in experiment
(Kahlweit et al., 1991). On the other hand, if $g(0) < 0$, then the condition
$g(0) < -g_{do}$ must be satisfied before the tricritical point is reached, indicating
that the lamellar phase is stable. In this case, a first-order transition to the
lamellar phase preempts the tricritical phase transition. Finally, there is the
possibility that $g(0)$ vanishes at the tricritical point, in which case it is a
Lifshitz tricritical point. This is the most interesting case, because the microemulsion retains its structure all the way to the Lifshitz tricritical point. From
the arguments given earlier in this section, the microemulsion should not wet
the oil/water interface right up to the Lifshitz tricritical point. To verify this,
Gompper et al. (1991) considered the symmetric tricritical point, $\phi_- = \phi_+$ in
(3.46). In this case, one can easily write the interfacial tension between phases
α and β calculated from (3.15) in the scaling form

$$\sigma_{\alpha\beta} = \omega \xi_L (\phi_+)^6 \hat{\sigma}_{\alpha\beta}, \tag{3.48}$$

where

$$\xi_L = [c/\omega(\phi_+)^4]^{1/4} \tag{3.49}$$

is the correlation length near the Lifshitz tricritical point. Note that *all*
tensions scale in the *same* way, contrary to the assumption of the Cahn
argument (Cahn, 1977). Because the contact angle for this symmetric case
is given by (Rowlinson and Widom, 1982)

$$\cos\theta = \frac{1}{2}\left(\frac{\sigma_{ow}}{\sigma_{om}}\right)^2 - 1,$$

$$= \frac{1}{2}\left(\frac{\hat{\sigma}_{ow}}{\hat{\sigma}_{om}}\right)^2 - 1, \tag{3.50}$$

one sees that, in the scaling regime, the contact angle is independent of the
distance to the Lifshitz tricritical point. The function $g(\phi)$ is written in the
scaling form $g(\phi) = c\xi_L^{-2}\hat{g}(\phi/\phi_+)$ which ensures the vanishing of $g(0)$ at the
tricritical point. The middle phase is a microemulsion for $|\hat{g}(0)| < 1$. The
particular form $\hat{g}(x) = 2\sqrt{2}x^2 - B$, with $|B| < 1$, is chosen, and the profiles
are again calculated from (3.17) and the scaled interfacial tensions are
evaluated from (3.15). The results show unequivocally that $\hat{\sigma}_{ow} < 2\hat{\sigma}_{om}$ so
that the oil/water interface is not wetted all the way to the Lifshitz tricritical
point.

3.3.3 Elastic properties of the amphiphile monolayer at the oil/water interface

One of the most striking features in self-assembling systems is the occurence of structures with *curved* monolayers, as in swollen spherical and cylindrical micelles. A calculation of the free energy of these structures allows contact to be made with the elastic theory of membranes (Canham, 1970; Helfrich, 1973; Evans, 1974) to be discussed in Section 4. In particular the spontaneous radius of curvature R_0, the bending rigidity κ of the monolayers, and the saddle-splay modulus $\bar{\kappa}$ can be determined. In terms of these parameters, the free energy of a curved surface reads for *small* local curvatures

$$\mathcal{H} = \int dS[\sigma + \lambda_s H + 2\kappa H^2 + \bar{\kappa}K] \tag{3.51}$$

where $H = (c_1 + c_2)/2$ is the mean curvature, and $K = c_1 c_2$ is the Gaussian curvature. Here, $c_i = 1/R_i$ ($i = 1, 2$), with the principal radii of curvature R_i, see Fig. 3.5. The unit surface element is dS, and the integration is over the whole interface. The spontaneous curvature modulus λ_s defines a spontaneous or preferred curvature $1/R_0$ according to $\lambda_s = -4\kappa/R_0$. For spheres, with $R_1 = R_2 = R$, the free energy per unit area is

$$f_{sphere} = \sigma + \frac{\lambda_s}{R} + \frac{2\kappa + \bar{\kappa}}{R^2} \tag{3.52}$$

while for cylinders, with $R_1 = R$, $R_2 = \infty$, the free energy per unit area is

$$f_{cyl.} = \sigma + \frac{\lambda_s}{2R} + \frac{\kappa}{2R^2}. \tag{3.53}$$

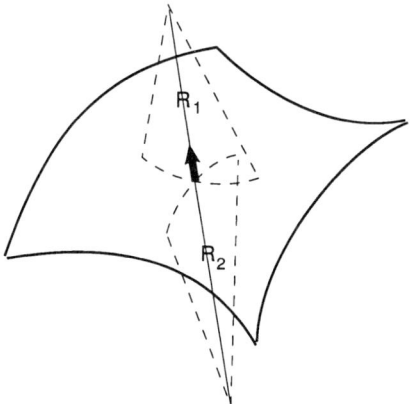

Fig. 3.5 The principal radii of curvature, R_1 and R_2, are the extremal values of the radius of curvature; they are also the eigenvalues of the curvature tensor.

Expressions for the elastic bending moduli have been derived from various microscopic models[†] and from Ginzburg–Landau models (Kawasaki and Kawakatsu, 1990; Kawakatsu and Kawasaki, 1990; Gompper and Zschocke, 1991, 1992b).

In Landau theory, the free energy of swollen spherical or cylindrical micelles of radius R is given by

$$\frac{F}{A} = \frac{1}{R^d} \int dr\, r^d \left[c\left(\frac{\partial^2}{\partial r^2}\phi_R(r) + \frac{d}{r}\frac{\partial}{\partial r}\phi_R(r)\right)^2 \right.$$

$$\left. + g(\phi_R)(\frac{\partial}{\partial r}\phi_R(r))^2 + f(\phi_R) \right] \tag{3.54}$$

where $d = 1$ (2) for the cylinder (sphere). Here, ϕ_R is the profile which extremizes the free energy (3.54), with the additional constraint that the radius be R. For large R, the interfacial profile ϕ_R approaches the planar profile, $\bar{\phi}(r - R)$. Therefore, the full profile can be expanded in powers of R^{-1}:

$$\phi_R(r) = \bar{\phi}(r - R) + \frac{\phi_1(r - R)}{R} + \frac{\phi_2(r - R)}{R^2} + \dots . \tag{3.55}$$

When only the first term in this expansion, the planar profile, is used, the first integral (3.36) of the Euler–Lagrange equation (3.35) can be used to eliminate f. Then, an expansion of the free energy in powers of R^{-1}, and identification of the coefficients of this expansion with the corresponding coefficients in (3.52) and (3.53), yields (Gompper and Zschocke, 1991, 1992b)

$$\sigma = \int_{-\infty}^{+\infty} dz\, p_s(z), \tag{3.56a}$$

$$\lambda_s = 2 \int_{-\infty}^{+\infty} dz\, z\, p_s(z), \tag{3.56b}$$

$$\kappa = \int_{-\infty}^{+\infty} dz\, 2c[\bar{\phi}']^2, \tag{3.56c}$$

$$\bar{\kappa} = \int_{-\infty}^{+\infty} dz\, \left(z^2\, p_s(z) - 4c[\bar{\phi}']^2\right), \tag{3.56d}$$

[†]The effect of electrostatic interactions on the curvature elasticity has been studied by Lekkerkerker (1989, 1990), Bensimon *et al.* (1990), Pincus *et al.* (1990), and Duplantier *et al.* (1990).

Estimates of the bending moduli in polymer systems have been discussed by Szleifer *et al.* (1988, 1990), Milner and Witten (1988), Wang and Safran (1990b, 1991) and Matsen and Schick (1993).

For one-component fluids such expressions have been derived by Blokhuis and Bedeaux (1991, 1992), Keller and Merchant (1991), Napiorkowski and Dietrich (1992, 1993) and Romero-Rochin *et al.* (1991, 1992).

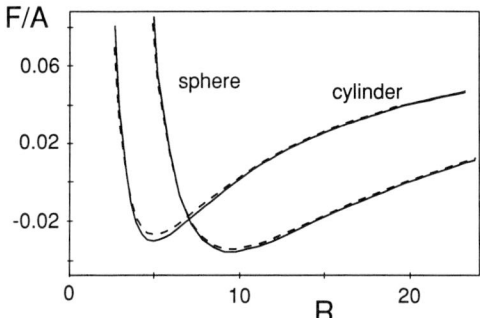

Fig. 3.6 The free energy of spherical and cylindrical micelles as a function of the radius R. The solid line is the Landau free energy (3.54) with $c = 1$, $g_+ = g_- = 4.6$, $g_0 = -4.5$, $\omega_+ = \omega_- = 4$, $\omega_0 = 1$, $\phi_+ = 2$, $\phi_- = -1$, $\mu = 0$, $f_0 = 1.4125$, the dashed line is the elastic bending energy (3.52) and (3.53) with the elastic constants calculated from (3.56). From Gompper and Zschocke (1991).

where

$$p_s(z) = 2g(\bar{\phi})[\bar{\phi}']^2 + 4c[\bar{\phi}'']^2 . \qquad (3.56e)$$

It should be noted that the interfacial tension obtained from the free-energy functional (3.15), with $\mu = 0$ and $f(\phi)$ eliminated from the invariant of (3.17), is the same as that obtained here. These expressions can be tested by calculating the free energy (3.54) explicitly for a particular choice of f and g, and then comparing it with (3.51) and (3.56). The result of such a calculation for the piecewise quadratic (constant) form of f (g) is shown in Fig. 3.6 for spheres and cylinders of radius R. The agreement of (3.54) with (3.51) and (3.56) is excellent. A different kind of check results from comparing the interfacial tension obtained from a lattice model within mean-field theory with that obtained from (3.56a) and (3.56e) and the Landau theory the lattice model generates. This has been done by Lerczak *et al.* (1992), and the comparison is shown in Fig. 2.15. One sees that the approximation is qualitatively correct for strong amphiphiles, and quantitatively correct for weak ones.

Several comments are in order:

(i) Note that the result (3.56) agrees with the results obtained for the standard square gradient theories, i.e. for $c = 0$, $g = \text{const} > 0$, in a low-temperature expansion (Diehl *et al.*, 1980; Lin and Lowe, 1983; Zia, 1985). In this case, $\kappa \equiv 0$, and $\bar{\kappa} > 0$. Equations (3.56c) and (3.56d) show that it is the presence of the Laplacian in the functional (3.34), which brings about a non-vanishing bending rigidity, and the possibility of a negative saddle splay modulus.

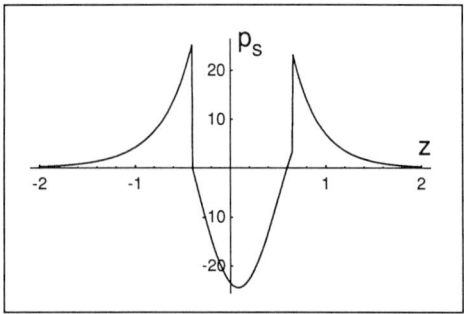

Fig. 3.7 Stress profile $p_s(z)$ through the monolayer at the oil/water interface (3.56e), for the piecewise parabolic model (3.21). The parameters are the same as in Fig. 3.6, and $f_0 = 1.42$. The asymmetry of p_s leads to a non-zero spontaneous curvature. From Gompper and Zschocke (1992b).

(ii) Helfrich (1981) has given a mechanical derivation of the elastic moduli. The surface tension, the spontaneous curvature modulus, and the saddle-splay modulus are found to be moments of the stress profile through the monolayer, while the bending rigidity vanishes identically. For $\kappa = 0$, the results (3.56) agree with this mechanical derivation, which leads to the identification of $p_s(z)$ as the stress profile. A typical stress profile for the piecewise parabolic model is shown in Fig. 3.7. Somewhat similar profiles have been obtained from molecular dynamics calculations of a molecular model (Smit *et al.*, 1990, 1991).

(iii) In the square gradient theories (metastable) mean-field solutions with spherical or cylindrical symmetry exist only for a pressure difference Δp between inside and outside, which satisfies the Laplace equation

$$\Delta p = \frac{(d-1)\,\sigma}{R}. \tag{3.57}$$

This is equivalent to a chemical potential difference μ between oil and water, which favours the interior phase (Rowlinson and Widom, 1982; Fisher and Wortis, 1984). Here, however, there can be another (stable) solution (with a smaller radius R) at oil–water coexistence. This additional solution of the mean-field equations exists only for $R \geq \xi$ (where ξ is the *bulk* correlation length of the interior phase), whereas it is always present when the Helfrich energy is minimized. Thus, the Hamiltonian (3.51) can only be valid for radii of curvature $R \geq \xi$.

(iv) The position of the interface has not been defined precisely yet. Inspection of (3.56) reveals that the surface tension σ and the bending rigidity κ are *independent* of the definition of the interface position. However, λ_s and the saddle splay modulus $\bar{\kappa}$ are not. For $\mu \neq 0$, the expression for λ_s agrees with those obtained by Rowlinson and Widom (1982) and Fisher and Wortis (1984), when the interface position is identified with the equimolar surface. However, for $\mu = 0$ we are free to define the interface position at will. It is easy to show that the free energy of a sphere or cylinder of fixed *physical* radius is independent of this definition.

It has been assumed thus far that the approximation of the full profile ϕ_R by the planar profile $\bar{\phi}$ can be used in (3.54) to calculate the elastic moduli in the Helfrich Hamiltonian (3.51). This is justified for the surface tension σ and the spontaneous curvature modulus λ_s, but is by no means obvious for κ and $\bar{\kappa}$. Therefore, it is essential to include the higher terms in the expansion (3.55) of the profile in the analysis. This has only been possible so far for droplets and cylinders stabilized by an internal pressure increment. In this case, the result of the calculation is an additional contribution to the free energy (Gompper and Zschocke, 1992b)

$$\delta F / A = -R^{-2} \int dr \left\{ \frac{1}{2} \mu_1 r \phi_1' + (d-1)[g(\bar{\phi})\bar{\phi}'\phi_1 + 2c\bar{\phi}''\phi_1'] \right\} \qquad (3.58)$$

where $d = 1$ (2) for the cylinder (sphere). Here, ϕ_1 satisfies the differential equation

$$2\phi_1'''' - 2g(\bar{\phi})\phi_1'' - 2g'(\bar{\phi})[\bar{\phi}''\phi_1 + \bar{\phi}'\phi_1'] - g''(\bar{\phi})[\bar{\phi}']^2\phi_1 + f''(\bar{\phi})\phi_1$$
$$= \mu_1 - 4(d-1)\bar{\phi}'' + 2(d-1)g(\bar{\phi})\bar{\phi}' \qquad (3.59)$$

and μ_1, the first term in the expansion of the chemical potential difference, $\mu = \mu_1/R + \mu_2/R^2 + ...$, is found to be

$$\mu_1 \int dr \bar{\phi}' = -(d-1)\, \sigma, \qquad (3.60)$$

which is just the Laplace equation (3.57). It was noted by Blokhuis and Bedeaux (1993) that from (3.59) and (3.60) follows $\phi_{1,\text{sphere}} = 2\phi_{1,\text{cyl}}$ and $\mu_{1,\text{sphere}} = 2\mu_{1,\text{cyl}}$. This can be used to extract κ and $\bar{\kappa}$ from (3.58). The result is that $\bar{\kappa}$ is unchanged while

$$\kappa = \int_{-\infty}^{\infty} dz \left[2c([\bar{\phi}']^2 - \bar{\phi}''\phi_{1,\text{sphere}}') - g(\phi)\bar{\phi}'\phi_{1,\text{sphere}} \right.$$
$$\left. - \frac{\mu_{1,\text{sphere}}}{4} z \phi_{1,\text{sphere}}' \right]. \qquad (3.61)$$

Unfortunately, the differential equation (3.59) for ϕ_1 cannot be easily solved. Analytic solutions have only been found for the case $c = 0$, $g(\phi) = g_0 = $ constant, and with a piecewise parabolic form for $f(\phi)$ (Gompper and Zschocke, 1992b), or with $f(\phi) = a_2\phi^2 + a_4\phi^4$ (Blokhuis and Bedeaux,

1993). In both cases κ changes from $\kappa = 0$ to $\kappa \simeq -2\bar{\kappa}$ when the ϕ_1-terms are taken into account. Thus for equilibrium droplets and cylinders stabilized by an internal pressure, the first-order correction to the curvature contribution of the free energy is larger than the zeroth-order term, so that the expression (3.56c) for κ is of little use in this case. For the more interesting case $\mu = 0$, the numerical results shown in Fig. 3.6 indicate that the expressions (3.56) give very good results for the elastic moduli, so that the curvature corrections seem not to be very important here.

3.3.4   Capillary waves

The elastic moduli calculated in the last section not only determine the shapes and sizes of micelles, but also the thermal fluctuations of the interface between oil and water. To see that this is so, the order-parameter distribution of the fluctuating interface is approximated by introducing a collective coordinate such that

$$\phi(\mathbf{r}) = \bar{\phi}(z - u(\mathbf{x})), \qquad\qquad \mathbf{r} = (\mathbf{x}, z), \qquad\qquad (3.62)$$

where $\bar{\phi}(z)$ is the order-parameter profile of the planar interface. By inserting this ansatz into the free-energy functional (3.15) with μ set to zero, and expanding to second order in the fluctuations $u(\mathbf{x})$, one finds (Gompper and Kraus, 1993a)

$$\mathcal{F}\{u\} = \sigma \int d^2x \left[1 + \frac{1}{2}(\nabla_\parallel u)^2\right] + \frac{1}{2}\kappa_0 \int d^2x \ (\nabla_\parallel^2 u)^2 + O(u^4) \qquad (3.63)$$

where σ and κ_0 are given by the integrals over the planar kink profile (3.56a) and (3.56c).

We now go beyond the collective coordinate approximation, and calculate the exact spectrum of Gaussian fluctuations of the interface (Zittartz, 1967; Gompper and Schick, 1990c, Gompper and Kraus, 1993a). For small fluctuations of the order-parameter field around the mean-field solution,

$$\phi(\mathbf{r}) = \bar{\phi}(z) + \eta(\mathbf{r}), \qquad\qquad (3.64)$$

the free-energy functional is expanded to second order in $\eta(\mathbf{r})$. Since the linear term in η vanishes due to the stationary property of $\bar{\phi}$, one finds

$$\mathcal{F}\{\phi\} = \mathcal{F}\{\bar{\phi}\} + \langle \eta, \hat{D}\eta \rangle, \qquad\qquad (3.65)$$

where $\langle,\rangle$ is the usual scalar product. The differential operator $\hat{D}$ can be written in the self-adjoint form (Gompper and Schick, 1990c)

$$\hat{D} = c\nabla^4 - g(\bar{\phi})\nabla^2 - g'(\bar{\phi})(\nabla\bar{\phi})\nabla - g'(\bar{\phi})\nabla^2\bar{\phi} - \frac{1}{2}g''(\bar{\phi})(\nabla^2\bar{\phi})^2 + \frac{1}{2}f''(\bar{\phi}) \ .$$

$$(3.66)$$

The fluctuations can now be expressed as a linear combination of eigenfunctions of $\hat{D}$,

$$\hat{D}\,\eta_\lambda(\mathbf{r}) = E_\lambda\,\eta_\lambda(\mathbf{r})\,. \tag{3.67}$$

Due to the translational symmetry of the mean-field solution parallel to the interface, the eigenmodes have the form

$$\eta_\lambda(\mathbf{r}) = \eta_{n\mathbf{q}}(z)\,e^{i\mathbf{q}\cdot\mathbf{x}} \tag{3.68}$$

so that $\eta(\mathbf{r}) = \sum_n \int d^2q\,\xi_{n\mathbf{q}}\,\eta_{n\mathbf{q}}(z)\,e^{i\mathbf{q}\cdot\mathbf{x}}$ which leads to

$$\langle\eta,\hat{D}\eta\rangle = \sum_n \int d^2q\,|\xi_{n\mathbf{q}}|^2\,E_{n\mathbf{q}}\langle\eta_{n\mathbf{q}},\eta_{n\mathbf{q}}\rangle \tag{3.69}$$

for the contribution of the fluctuations to the free energy. Here, the index n stands for both the discrete and the continuous parts of the spectrum.

Equation (3.67) has a solution $\eta_{00}(z) = \bar{\phi}'(z)$ with eigenvalue $E_0 = 0$, because it costs no energy to displace the interface in the z-direction. We are interested in the wavevector dependence of the translational mode $\eta_{0\mathbf{q}}(z)$, which is the lowest energy mode for $q = 0$.

To obtain the coefficients of the q^2- and q^4-contributions to the spectrum, one expands

$$\eta_{n\mathbf{q}}(z) = \eta_n^{(0)}(z) + q^2\eta_n^{(2)}(z) + q^4\eta_n^{(4)}(z) + O(q^6), \tag{3.70}$$

substitutes this into the eigenvalue equation (3.67), and calculates the q-dependent contribution to $\langle\eta,\hat{D}\eta\rangle$ from the lowest mode $\eta_{0\mathbf{q}}(z)$. One obtains (Gompper and Kraus, 1993a)

$$E_{0\mathbf{q}}\langle\eta_{0\mathbf{q}},\eta_{0\mathbf{q}}\rangle = \frac{1}{2}\sigma_t q^2 + \frac{1}{2}\kappa q^4 + O(q^6)\,, \tag{3.71}$$

with

$$\sigma_t = \langle\eta_0^{(0)},\hat{\Gamma}\eta_0^{(0)}\rangle \tag{3.72}$$

$$\kappa = \kappa_0 + 2\langle\eta_0^{(2)},\hat{\Gamma}\eta_0^{(0)}\rangle\,, \tag{3.73}$$

and the operator

$$\hat{\Gamma} = g(\bar{\phi}(z)) - 2c\nabla_z^2\,. \tag{3.74}$$

For a free interface σ_t, which is the coefficient of the q^2 term of the spectrum (3.71), is identical to the mean-field expression for the interfacial tension, (3.15) with $\mu = 0$, or (3.56a). This is expected. However, κ, which is the coefficient of the q^4 term in the spectrum, differs from the mean-field value κ_0 by a contribution arising from the deviation of the curved profile from the form given by (3.62).

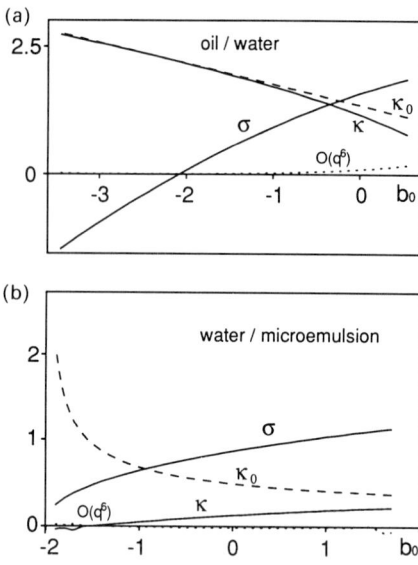

Fig. 3.8 Interfacial tension σ and bending rigidity κ of the interface between two homogeneous phases at oil–microemulsion–water coexistence, as calculated from the spectrum of capillary waves, with $g_+ = g_- = 4.5$, $f_0 = 0$, and $c = 1$, $\omega_0 = 1$, $\omega_+ = \omega_- = 4$, $\phi_+ = \phi_- = 1$, $\mu = 0$. (a) Oil/water interface. (b) Water/microemulsion interface. The tension σ is positive at oil–water coexistence; when $\sigma < 0$, the lamellar phase is stable. Note that κ, which takes into account the curvature-induced change of the profile, can differ significantly from κ_0 (dashed lines). Terms of order q^6 (dotted lines) in the spectrum are very small, and contribute only for the water/microemulsion interface near $g_0 = -2$, where κ is negative, so that the excitation energy is always positive. From Gompper and Kraus (1993a).

The results for the piecewise parabolic model (3.21) are shown in Fig. 3.8(a) for the oil/water interface, and in Fig. 3.8(b) for the oil/microemulsion interface. We see that for the *oil/water* interface, the approximation of the local order-parameter profile by the mean-field profile gives very good results for $g_0 \lesssim -1$, i.e. in a region of the phase diagram, where the correlation function in the microemulsion has strong oscillations, so that the interface can be considered to be covered by a saturated amphiphilic monolayer. For larger values of g_0, the deviation of the profile of the curved interface from the simple form (3.62) becomes more and more important. Of course, as the system approaches the wetting transition, which occurs at $g_0 = 2$, the width of the interface increases. At the transition, a second zero-mode appears in the spectrum which describes fluctuations of the interface width.

For the *oil/microemulsion* interface, the simple approximation (3.62) always fails to give adequate results. Thus we conclude that this interface is a more complicated object, with an interfacial profile which is strongly distorted when the interface is bent. Its description by means of a curvature model breaks down completely near $g_0 = -2$. Here the term of order q^6 ensures the stability of the interface.

The result (3.71) for the spectrum of capillary waves also applies to fluctuations of oil/water interfaces in the *lamellar phase*. It is well known that the undulation modes destroy the long-range order in lamellar phases (Peierls, 1934; Landau, 1965). For spatially modulated phases, the free-energy density is not only minimized with respect to the order-parameter profile, but also with respect to the periodicity length. It can be shown (Gompper and Kraus, 1993a) that this implies

$$\int_0^{z_1} dz \left[g(\bar{\phi})(\bar{\phi}')^2 + 2c(\bar{\phi}'')^2 \right] = \int_0^{z_1} dz \; \bar{\phi}' \hat{\Gamma} \bar{\phi}' = 0 \qquad (3.75)$$

for undulation modes, for which all interfaces fluctuate in phase. Thus σ_t, (3.72), which appears in the quadratic part of the spectrum, vanishes identically in this case. The leading term in the spectrum is therefore controlled by the bending energy. We emphasize that the absence of q^2-contributions does not imply that the interfacial free energy of a single oil/water interface is identically zero in the lamellar phase. In fact, the calculated interfacial free energy is generally non-zero, and usually negative in the region of the phase diagram in which the lamellar phase is stable. Of course, it must be positive at three-phase coexistence of oil, water and lamellar phases. Just as the calculated oil/water interfacial free energy is not zero in the lamellar phase, it is almost certainly not zero where the microemulsion is stable.

For very long wavelengths, the effect of van der Waals interactions cannot be ignored. In calculating the elastic properties of an interface in a one-component system, Napiorkowski and Dietrich (1992, 1993) find that it gives rise to a $q^4 \ln(q)$ contribution to the capillary spectrum. This is an effect which is beyond the range of validity of the curvature Hamiltonian (3.51).

3.3.5 Phase diagram

The phase diagram of the Landau model (3.15) with the functions $f(\phi)$ and $g(\phi)$ given by (3.21) contains three homogeneous phases by construction. There are eleven parameters, but by rescaling $\mathbf{r}$, ϕ, and the free energy itself, only eight are independent. No attempts have been made so far to map out the whole parameter space. The most interesting case is certainly that in which the properties of oil and water are fixed (and their chemical potential difference is zero), and the amphiphile is varied in concentration, reflected in f_0, and strength, g_0. This situation was studied by Gompper and Zschocke (1992b). For $g_0 \gtrsim - g_{do}$, the homogeneous phases are stable within mean-field

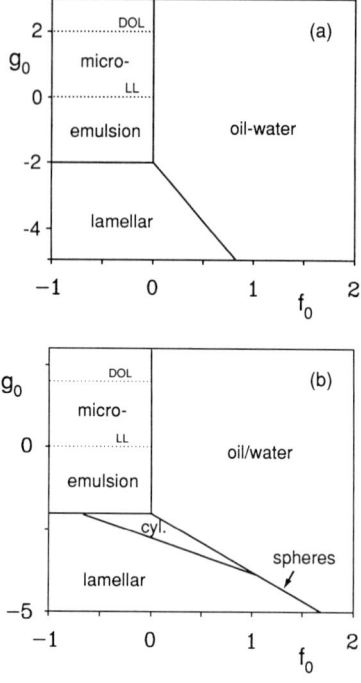

Fig. 3.9 Mean-field phase diagrams in the space of f_0 and g_0 with $\mu = 0$. The parameters in (3.15) and (3.21) are $c = 1$, $g_+ = g_- = 4.6$, $w_+ = w_- = 4$, $w_0 = 1$. All transitions are first order. The correlation function changes its behaviour at the disorder line (DOL) from oscillatory to monotonic. The peak of the scattering intensity moves away from $q = 0$ at the Lifshitz line (LL). (a) System with oil–water symmetry, $\phi_+ = 1$, $\phi_- = -1$. (b) System with broken oil–water symmetry, $\phi_+ = 2$, $\phi_- = -1$. From Gompper and Zschocke (1992b).

theory; the microemulsion for $f_0 < 0$, the oil- and water-rich phases otherwise. For $g_0 < -g_{do}$, spatially modulated phases are stable. Two phase diagrams are shown in Fig. 3.9, the first for the case of oil–water symmetry, the second without this symmetry. In the former case, the spontaneous curvature modulus vanishes by symmetry, so that only a lamellar phase is expected. That is indeed what is found. In the latter case, three different modulated phases occur:

(1) a lamellar phase;
(2) a hexagonal phase of oil cylinders in water (or *vice versa*) of which the order-parameter distribution is displayed in Fig. 3.10;
(3) a cubic crystal of oil droplets (swollen micelles) in water (or *vice versa*).

Fig. 3.10 Contour plot of the order-parameter density in the hexagonal phase of cylinders. The parameters of (3.21) are the same as in Fig. 3.9(b), and $f_0 = 0.808$, $g_0 = -3.5$. From Gompper and Zschocke (1992b).

For the ϕ^6-potential (3.19), (3.20), metastable bicontinuous cubic phases in which both oil- and water-channels traverse the whole system, are also found. As the phase transition from the lamellar phase to the microemulsion is approached, the bicontinuous ordered phases become degenerate with the lamellar phase, which indicates that the microemulsion itself is bicontinuous.

It is instructive to compare the phase diagram, Fig. 3.9(b), of the non-symmetric system with the results for the elastic moduli (3.56), for the same system. These moduli are shown in Fig. 3.11. Note that the transition from oil–water coexistence to the spatially modulated phases occurs very close to the line where the interfacial tension of the oil/water interface vanishes. As argued earlier in Section 2.2.3, the deviation of the line of phase transitions from the line $\sigma = 0$ is a measure for the magnitude of the attractive interaction between monolayers. Because the spontaneous curvature is non-zero, one expects, and finds, cubic phases of swollen micelles as well as hexagonal phases of close-packed cylinders. The saddle-splay modulus is negative over a large part of the $\sigma = 0$ line, which indicates the existence of a droplet phase, again in agreement with the calculated phase diagram. This phase is also favoured below the $\sigma = 0$ line by the negative saddle-splay modulus $\bar{\kappa}$. However as the concentration of amphiphile increases, with small or negative f_0 and large and negative g_0, the packing constraints always favour the lamellar phase.

Elastic moduli have also been calculated from the Landau theory generated by the three-component model of Section 2.1.1. One finds in the symmetric case that along the line of three-phase coexistence between oil-rich, water-rich and middle phase the saddle-splay modulus is positive; the bending rigidity increases as the lamellar phase is approached (Lerczak *et al.*, 1992).

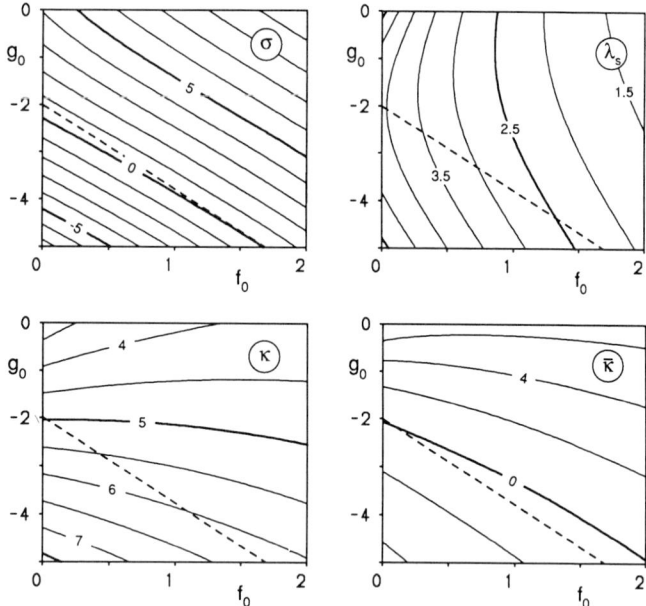

Fig. 3.11 Contour plot of surface tension σ, spontaneous curvature modulus λ_s, bending rigidity κ, and saddle-splay modulus $\bar{\kappa}$, as a function of f_0 and g_0. The other parameters in (3.21) are the same as in Fig. 3.9(b). The dashed line marks the position of the phase transition between oil–water coexistence and the spatially modulated phases. From Gompper and Zschocke (1991).

3.3.6 Monte Carlo simulations

Monte Carlo simulations can be carried out on the one-component Ginzburg–Landau model of amphiphilic systems (3.15). To simulate a continuum Ginzburg–Landau model, space is discretized, so that it is in fact a lattice model which is simulated. However, there are two important differences between this and the lattice models discussed in Section 2: (i) here, the variables on the lattice sites are continuous; (ii) the lattice constant has no physical meaning, and can be chosen such that lattice effects are minimized.

The system has been simulated by Gompper and Kraus (1993b) for $\mu = 0$, $c = 1$, and the following choice of $f(\phi)$ and $g(\phi)$

$$f(\phi) = (\phi^2 - 1)^2(\phi^2 + f_0), \tag{3.76}$$

$$g(\phi) = g(0) + g_2\phi^2, \tag{3.77}$$

with

$$g_2 = 4\sqrt{1 + f_0} - g(0) + 0.01. \tag{3.78}$$

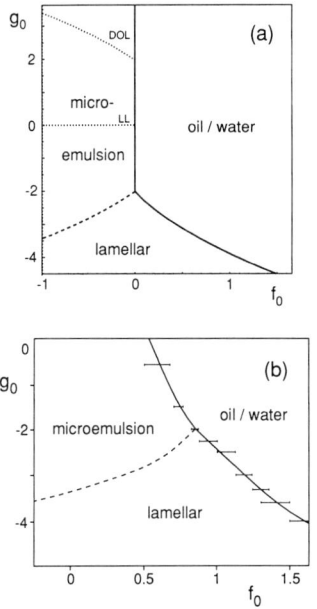

Fig. 3.12 (a) Mean-field phase diagram of the Ginzburg–Landau model with $f(\phi)$ and $g(\phi)$ given by (3.76), (3.77). The Lifshitz (LL) and disorder (DOL) lines are shown dotted. (b) Phase diagram as obtained by Monte Carlo simulation. Error bars denote regions where metastable phases occur. The weak first-order transition from the disordered to lamellar phase is shown with a dashed line. From Gompper and Kraus (1993b).

The bulk free-energy density, $f(\phi)$ has minima of equal depth at $\phi = \pm 1$, and a third minima of adjustable height at $\phi = 0$. Three-phase coexistence occurs at $f_0 = 0$. The strength of the amphiphile is varied by $g(0)$. Irrespective of the choice of $g(0)$, the correlation functions in the oil- and water-rich phases decay monotonically at large distances. The mean-field phase diagram of this model is shown in Fig. 3.12(a). Clearly it is very similar to that with piecewise continuous $f(\phi)$ and $g(\phi)$ shown in Fig. 3.9(a).

For comparison, the phase diagram obtained from simulations is shown in Fig. 3.12(b). The phase transition between the microemulsion and the lamellar phase is found to be very weakly first order, in agreement with the general result that a transition to a phase characterized by a finite wavevector should be fluctuation-induced first order (Brazovskii, 1975). The transitions from oil–water coexistence to both the microemulsion and the lamellar phase are strongly first order; in this case, hysteresis effects make a precise determination of the transition line difficult.

A comparison of the Monte Carlo and the mean-field phase diagrams shows that the region of stability of the microemulsion increases due to the fluctuations, both towards the oil–water coexistence and the lamellar phase. Also the lamellar phase extends into regions where the oil-rich or water-rich phases are stable in mean-field theory. Thus, the fluctuations stabilize phases with an extensive amount of internal interface.

The shift of the transitions from oil–water coexistence to a phase with internal interfaces can be understood by calculating the renormalization of the oil/water interfacial tension due to fluctuations (Gompper and Kraus, 1993b). The renormalized tension is the free energy per unit area, which can be obtained via a functional integration over $\exp(-\mathcal{F}\{u\})$, where $\mathcal{F}\{u\}$ is the capillary wave free-energy functional of (3.63). The result of this functional integration is $\exp(-\sigma_R \mathcal{A})$, with $\mathcal{A}$ the area of the interface in the simulation, and σ_R the interfacial tension which includes Gaussian fluctuations. The line $\sigma_R = 0$ is found to reproduce quite well the shape of the line of phase transitions from the disordered phase to oil–water coexistence. Similarly, the phase transition from the microemulsion to the lamellar phase can be understood by studying the bending rigidity κ of the interfaces. In this case, renormalization effects (Peliti and Leibler, 1985; Helfrich, 1985) are not important, because they only change κ by an additive constant. The line $\kappa = 2.5 k_B T$ is found to describe the shape of the transition line reasonably well. This makes contact with the membrane theories of Section 4 which, as we will see, also predict the transition to the lamellar phase to occur at constant bending rigidity.

The simulations yield structural information on the microemulsion phase which goes considerably beyond what can be extracted from the mean-field and Ornstein–Zernike approximations. The most interesting part of the phase diagram is the region near four-phase coexistence. A typical configuration of the microemulsion phase is shown in Fig. 3.13. A cut through this structure strongly resembles the pictures of microemulsion structure obtained experimentally by freeze-fracture microscopy (Jahn and Strey, 1988).

The structure factor of the Ginzburg–Landau model calculated from the Monte Carlo simulations is different from the Ornstein–Zernike result. Nevertheless, the expansion of the scattering intensity for small wavevector q to fourth order in q is still possible,

$$S(q) = \frac{1}{\tilde{a}_2 + \tilde{g}_0 q^2 + \tilde{c}q^4}, \tag{3.79}$$

and is a convenient way of parameterizing the function, just as in the analysis of experimental data (Teubner and Strey, 1987; S.-H. Chen et al., 1990, 1991). By Fourier transformation of this scattering intensity, the correlation length ξ and the characteristic wavelength λ of the asymptotic decay of the correlation function can be extracted, and are given by (3.13) and (3.14). As noted earlier,

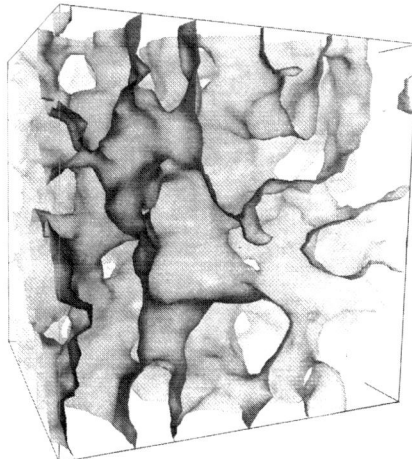

Fig. 3.13 Typical equilibrium configuration in the microemulsion in the vicinity of the transition to the lamellar phase, at $g(0) = -2.5, f_0 = 0.675$. Only the $\phi(\mathbf{r}) = 0$ surfaces are shown. From Gompper and Kraus (1993b).

the ratio $2\pi\xi/\lambda$ takes the values 0 and 1 at the disorder and Lifshitz lines respectively, and diverges as a continuous transition to a lamellar phase is approached. The results obtained from the simulation for this ratio are shown in Fig. 3.14.

These values can be compared with experimental ones (Teubner and Strey, 1987; Widom, 1989; S.-H. Chen *et al.*, 1990, 1991; Schubert and Strey, 1991)

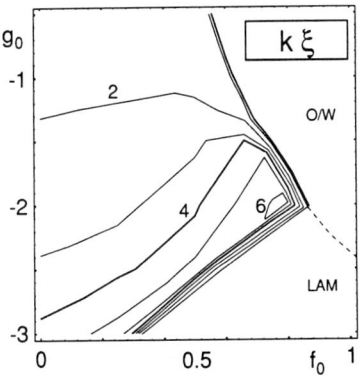

Fig. 3.14 The dimensionless product $2\pi\xi/\lambda$, which is the ratio of the two length scales in the correlation function. The dashed line indicates a portion of the phase boundary between the lamellar phase and oil–water coexistence. From Gompper and Kraus (1993b).

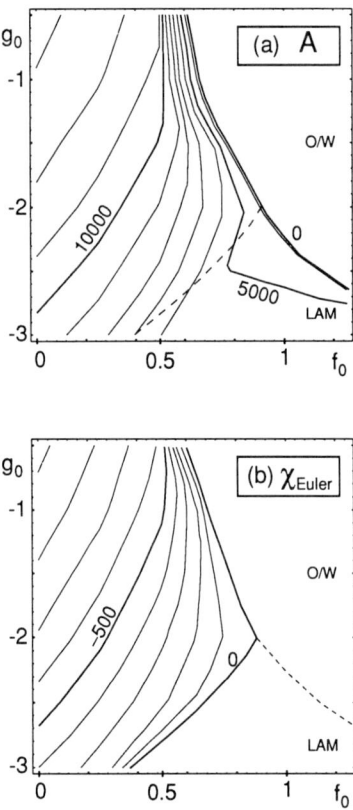

Fig. 3.15 Quantities characterizing the structure of the microemulsion: (a) area A of internal oil/water interface; (b) Euler characteristic χ_E. The dashed lines indicate the positions of the phase transitions. From Gompper and Kraus (1993b).

which are in the range from 0 to 4.77, with higher values for systems with large amphiphile concentrations, i.e. near the lamellar phase. This is in good agreement with the results presented in Fig. 3.14.

The microemulsion can also be characterized by its Euler characteristic χ_E and its internal area A, here defined by the surface $\phi = 0$. Contour plots of A and χ_E are shown in Figs. 3.15(a) and 3.15(b), respectively. A negative Euler characteristic in the microemulsion phase indicates a bicontinuous structure with multiply connected labyrinths of oil-rich and water-rich regions. The internal area in the microemulsion decreases with decreasing distance to the lamellar phase transition. Experimentally, one approaches the lamellar transition by increasing the concentration of amphiphile. Therefore, the simulation indicates that the internal area in the microemulsion actually decreases with *increasing* concentration of amphiphile. This is precisely the opposite of the behaviour which would occur if all internal interfaces were completely covered with amphiphile. Thus the data can be interpreted as showing that the internal interfaces are, in general, not completely coated. This is in agreement with simulations of the three-component lattice model, Fig. 2.8, and the simulations of the Larson model, Fig. 2.27. As the lamellar transition is approached, the internal interfaces become more and more saturated.

In a bicontinuous structure, the area and the Euler characteristic are not independent. Let us assume that the microemulsion of total volume $V = L^3$ is composed of building blocks of typical length scale L_0. All these elementary units have approximately the same structure and topology, with a typical Euler characteristic χ_0, which is independent of L_0. Then, the Euler characteristic of the whole system is

$$\chi_E(V) = \chi_0 \left(\frac{L}{L_0}\right)^3 . \tag{3.80}$$

The internal area of the elementary unit is $A(V_0) = \tilde{A}L_0^2$, so that the area of the whole system is given by

$$A(V) = \tilde{A}L_0^2 \left(\frac{L}{L_0}\right)^3 = \tilde{A}V^{2/3}\frac{L}{L_0} . \tag{3.81}$$

By eliminating L/L_0 from these equations, one obtains

$$\sqrt[3]{-\chi_E(V)} = \sqrt[3]{-\chi_0 V^{-2}\tilde{A}^{-1}} \, A(V) . \tag{3.82}$$

The Monte Carlo results indeed show this scaling behaviour. Furthermore, the data imply $\sqrt[3]{-\chi_0}/\tilde{A} \approx 0.56$. This value characterizes the structure of the microemulsion within an elementary unit. It can be compared with the value of $\sqrt[3]{-\chi_0}/\tilde{A}$ for *ordered* bicontinuous minimal surfaces (i.e. surfaces with zero mean curvature) (Anderson *et al.*, 1990). One finds $\sqrt[3]{-\chi_0}/\tilde{A} \approx 0.680$ for the Schwarz P-surface, 0.657 for the D-surface, and values in the range 0.66–0.72 for more complicated structures. The difference between these values and that for the microemulsion results from the fact that the microemulsion is not an ordered phase, and that it also contains a small number of micelles with positive Euler characteristic and negligible area.

The amphiphilic monolayers in the lamellar phase show strong undulations due to thermal fluctuations. These fluctuations have been studied intensively within the curvature model (3.51), (3.63) for membranes (Helfrich, 1978, 1985). In the lamellar phase, the fluctuations $u(\mathbf{x})$ are restricted due to the presence of other membranes, so that $\langle u^2 \rangle = \nu d^2$, where d is the distance between membranes, and ν is a constant of order unity. When the fluctuations of the monolayers are neglected, the internal area of the lamellar phase is obtained from a simple geometrical relation, $A = Vd^{-1}$. The inclusion of fluctuations increases the internal area. This excess area depends on the separation of the membranes. For small undulations, the excess area is found to be (Helfrich, 1985)

$$\frac{\Delta A}{A} = \frac{k_B T}{8\pi\kappa} \ln \frac{\sigma/\kappa + q_{max}^2}{\sigma/\kappa + q_{min}^2} \tag{3.83}$$

(for a more detailed discussion see Section 4.3 below). Here, the high momentum cutoff q_{max} is determined by the intrinsic width of the oil/water interface, $q_{max} = \alpha 2\pi/\xi_{kink}$, with a constant α of order unity. The low momentum cutoff is $q_{min} = 2\pi/\xi_\parallel$, where $\xi_\parallel$ is the parallel correlation length. It is determined by the requirement $\langle u^2 \rangle = \nu d^2$. Finally, the distance d is the spacing available for the fluctuations of a membrane. Therefore, the width of a single interface has to be subtracted from the periodicity length of the lamellar phase,

$$d = \frac{L}{n} - 2\xi_{kink} , \tag{3.84}$$

where n is the number of interfaces in a system of length L. In the case $\sigma = 0$, the excess area is then given by

$$\frac{\Delta A}{A} = \frac{k_B T}{4\pi\kappa} \ln \left(\frac{\sqrt{4\pi\kappa\nu}}{\xi_{kink}} \alpha d \right) . \tag{3.85}$$

This logarithmic behaviour of the excess area with membrane separation has been confirmed experimentally (Roux et al., 1992b) for strongly swollen lamellar phases.

In the Monte Carlo simulations, the limit where all other length scales are small compared with the interface separation cannot be reached, due to the finite system size. Furthermore, as noted earlier, the interfacial tension σ only vanishes for the equilibrium distance of the lamellar phase. Due to the finite system size, this optimal distance usually cannot be attained. For all other separations, exponential corrections to the interfacial tension have to be taken into account (Gompper and Kraus, 1993a,b),

$$\sigma = \sigma_\infty + \bar{\sigma} e^{-d/\xi_{bulk}} , \tag{3.86}$$

where σ_∞ is the renormalized interfacial tension.

The excess area for a point in the phase diagram near the four-phase point is shown in Fig. 3.16 as a function of the inverse interface separation $1/d$. The data are compared with the logarithmic behaviour (3.85), which fits the data only over a small interval. However, when the exponential distance dependence of the interfacial tension (3.86), is taken into account in (3.83),

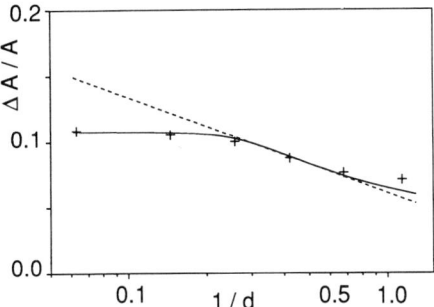

Fig. 3.16 Excess area in the lamellar phase at $g(0) = -2.25$, $f_0 = 1.125$ for various monolayer separations. Error bars are of the order of the symbol size. The dashed line is the theoretical prediction with $\sigma = 0$ (3.85); the solid line is obtained using (3.83) and (3.86) with $\sigma_\infty = 0.25$. The parameters $\nu = 0.16$, $\bar{\sigma} = 12$ and $\alpha = 1.6$ have been used. From Gompper and Kraus (1993b).

all data can be explained very well. Thus, the existence of an exponential, d-dependent contribution in the interfacial tension is clearly confirmed by the simulation results. Such a term has also been found to be important in the context of wetting transitions (Fisher and Jin, 1992).

3.4 Aqueous amphiphilic solutions

3.4.1 The L_3 phase and inside/outside transitions

In the system of water and amphiphile, the latter can form bilayer sheets which play a role similar to the monolayers in ternary water–oil–amphiphile systems. Thus one can expect in the binary system an analogue of the microemulsion in the ternary one. This analogue is denoted the L_3, or "sponge" phase. The question arises as to how one should describe it theoretically, and observe it experimentally. One naturally considers the kind of signal which would arise from scattering experiments. If the phase were treated as an ordinary fluid, one would expect the amphiphiles in solution to give rise to a normal Lorentzian structure function. This would ignore the contribution arising from the fluctuations of the bilayers, and it is this contribution that we wish to include. To address this, Cates *et al.* (1988a,b) argued as follows. In the ternary system, the monolayers divide distinct regions of oil and water, and the concentration difference of these two components is one of the order parameters used to construct a theory of these ternary mixtures. Similarly, the bilayers divide the water into distinct regions, which can be labelled "inside" and "outside". Just as the fluctuations in the order parameter in the ternary system give rise to a correlation and associated structure function, so do they in the binary system. The major difference is that the fluctuations of the order parameter can be observed directly in the ternary system because the incident particles are sensitive to the density difference between oil and water, whereas they cannot be observed directly in the binary system because the incident particles are not sensitive to the difference between inside and outside water. None the less, the fluctuations in the inside/outside order parameter do couple to the fluctuations in the amphiphile density, and therefore are observable in the amphiphile structure function. As the form of the resulting structure function differs from a Lorentzian, one has an experimental marker for the presence of fluctuating bilayers and the L_3 phase.

To make these ideas concrete, we consider a Landau theory with two order parameters (Roux *et al.*, 1990, 1992a); ϕ, the local difference between inside and outside water concentrations, and ψ the local concentration of amphiphile. We select from (3.22) and (3.23) the following terms

$$
\begin{aligned}
\mathcal{F}_0[\phi, \psi] = \int \mathrm{d}^3 r [\beta_1 (\nabla \phi)^2 + \tilde{A}_2 \phi^2 + \tilde{A}_4 \phi^4 \\
+ \delta_1 (\nabla \psi)^2 + \tilde{B}_2 \psi^2 - \mu_s \psi \\
+ \gamma_1 \phi^2 \nabla^2 \psi + \tilde{C}_3 \phi^2 \psi].
\end{aligned}
\tag{3.87}
$$

The analysis is simplified for the case in which the coefficients δ_1 and γ_1 are small enough to be ignored. In that case, the fluctuations of ϕ are of the Ornstein–Zernike form

$$
\langle \phi(\mathbf{r})\phi(0)\rangle = \frac{\exp(-r/\xi)}{8\pi\beta_1 r},
\tag{3.88}
$$

where the correlation length

$$
\xi = (\beta_1/\tilde{A}_2)^{1/2}.
\tag{3.89}
$$

Again, this structure function is not observable because of the lack of contrast between inside and outside. One now considers the amphiphile correlation function $\langle \psi(\mathbf{r})\psi(0)\rangle$. From (3.87) with $\delta_1 = \gamma_1 = 0$, one sees that the probability distribution for $\psi(\mathbf{r})$ is Gaussian with variance $1/2\tilde{B}_2$. Therefore

$$
\langle \psi(\mathbf{r})\psi(0)\rangle = \frac{\tilde{C}_3^2}{4\tilde{B}_2^2}[\langle \phi^2(\mathbf{r})\phi^2(0)\rangle - \langle \phi^2 \rangle^2] + \frac{\delta(\mathbf{r})}{2\tilde{B}_2}.
\tag{3.90}
$$

Under the assumption that the fluctuations of ϕ are Gaussian, the value of the square bracket is simply $2\langle\phi(\mathbf{r})\phi(0)\rangle^2$ which we know from (3.88). Hence

$$\langle\psi(\mathbf{r})\psi(0)\rangle = \frac{\tilde{C}_3^2}{2\tilde{B}_2^2}\frac{\exp(-2r/\xi)}{(8\pi\beta_1 r)^2} + \frac{\delta(r)}{2\tilde{B}_2}. \tag{3.91}$$

The Fourier transform of this function is the structure function of experimental interest. It is given by (Roux *et al.*, 1990)

$$S(\mathbf{q}) = \frac{1}{2\tilde{B}_2} + \frac{\tilde{C}_3^2\pi\xi}{(8\pi\beta_1\tilde{B}_2)^2}\frac{\tan^{-1}(q\xi/2)}{(q\xi/2)}. \tag{3.92}$$

This result should apply for q^{-1} much larger than the distance between bilayers, a region in which the correlations between amphiphiles should be dominated by the fluctuations in the bilayers. It should not apply for such small q^{-1} at which scattering would probe the structure of individual bilayers. For $q\xi$ large, $S(q)$ for the system with many internal bilayers falls like $1/q$, rather than the $1/q^2$ of a normal fluid without the internal bilayers. When the couplings δ_1 and γ_1 are restored, there are quantitative but no qualitative changes (Roux *et al.*, 1992a). One now finds the usual Lorentzian contribution from the amphiphiles in solution which falls as $1/q^2$, but this is dominated by the $1/q$ contribution which remains.

A candidate for an L_3 phase occurs in the system of water, ionic amphiphile AOT, and salt (Skouri *et al.*, 1991). There is a narrow region of isotropic phase which exists quite close to a lamellar phase, just as the microemulsion is found close to a lamellar phase. The isotropic phase can exist down to amphiphile volume fractions as low as about 0.05. Neutron scattering results for three different concentrations of amphiphile are shown in Fig. 3.17. There is a small shoulder at $q \approx 0.01\,\text{Å}^{-1}$ which indicates a characteristic distance between sheets of 600 Å. Data from light scattering on a sample with amphiphile volume fraction $\psi = 0.0432$ are shown in Fig. 3.18 together with the best fit to the data using (3.92). The agreement is excellent. Note however the very restricted range of wavevectors, only about one third of the distance to the peak.

One means of going beyond the above analysis is to exploit the analogy with the microemulsion (Gompper and Schick, 1994b). There we found that the coefficient of the square gradient term was negative, reflecting the fact that one effect of the amphiphile is that gradients in the order parameter are energetically favourable. This would entail taking the coefficient β_1 to be negative in the functional of (3.87) and adding a term $\beta_2(\nabla^2\phi)^2$ with $\beta_2 > 0$ for stability. This now leads to the correlation function found previously, (3.12),

$$\langle\phi(\mathbf{r})\phi(0)\rangle = \frac{\lambda\xi}{32\pi^2\beta_2 r}e^{-r/\xi}\sin\frac{2\pi r}{\lambda}, \tag{3.93}$$

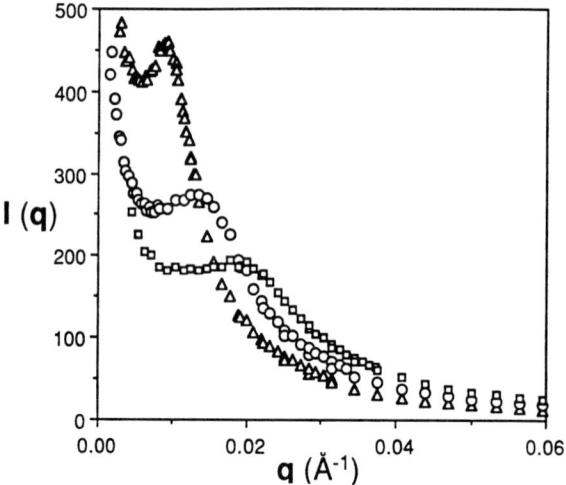

Fig. 3.17 Neutron scattering patterns of L_3 samples in the ternary system with volume fractions $\psi = 0.0474$ (triangles), 0.0675 (circles) and 0.0897 (squares). From Skouri *et al.* (1991).

with λ and ξ given by (3.13) and (3.14) with a_2, g_0 and c replaced by $\tilde{A}_2$, β_1 and β_2. By repeating the calculation that led to (3.92), that is by setting $\gamma_1 = \delta_1 = 0$, one readily obtains

$$S(\mathbf{q}) = \frac{1}{2\tilde{B}_2} + \frac{\tilde{C}_3^2 \pi \xi}{\tilde{B}_2^2} \left(\frac{\lambda \xi}{32\pi^2 \beta_2}\right)^2 \frac{1}{q\xi/2} \left[\frac{1}{2} \tan^{-1}\left(\frac{q\xi}{2}\right)\right.$$

$$\left. - \frac{1}{4} \tan^{-1}\left(\frac{(2k+q)\xi}{2}\right) + \frac{1}{4} \tan^{-1}\left(\frac{(2k-q)\xi}{2}\right)\right], \qquad (3.94)$$

where $k \equiv 2\pi/\lambda$, the characteristic wavevector of the bilayer structure. One

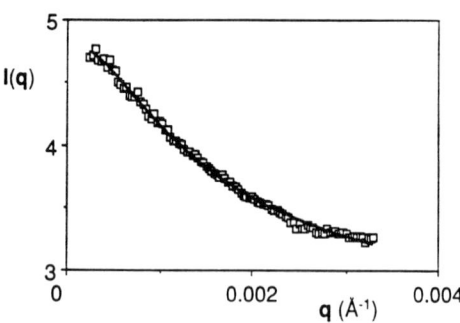

Fig. 3.18 Fit of the light scattering data from the L_3 phase with $\psi = 0.0432$ to the form given in (3.92). From Skouri *et al.* (1991).

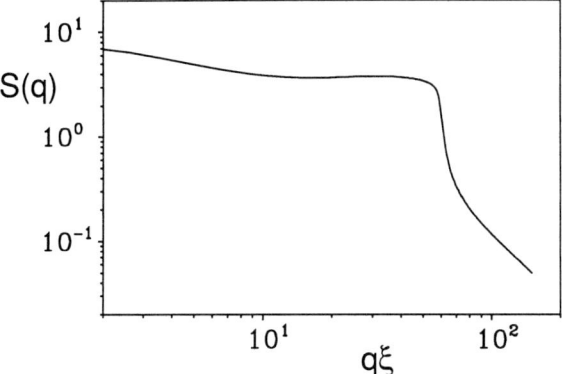

Fig. 3.19 Form of the structure function for the amphiphile density assuming that the order-parameter correlation function is of the form of (3.93).

easily sees from this that for $q \ll 2k$, the second and third terms cancel leaving the same result (up to a scale factor of $1/2$) as before (3.92), assuring the same excellent fit at small wavevectors. For $q \gg 2k$, however, the last two terms almost cancel the first, causing $S(q)$ to drop at such values and fall as $1/q^3$. Again the coupling terms proportional to δ_1 and γ_1 in (3.87) can be restored leading to the presence of an ordinary Lorentzian characterized by the usual bulk correlation length. However in contrast to the previous case, this makes a qualitative difference at large wavevector as the Lorentzian falls as $1/q^2$ and eventually dominates the $1/q^3$ contribution. Thus one finds three different behaviours as a function of wavevector: the initial fall with $1/q$ obtained in experiment, then a crossover to $1/q^3$ behaviour, and finally at sufficiently large wavevector, a $1/q^2$ dependence. There is a peak in $S(q)$ between the first two regions. A typical result is shown in Fig. 3.19, which should be compared with Fig. 3.17. When the additional amphiphile coupling γ_2 of Eq. (3.23) is included, the agreement with experiment becomes quite good (Gompper and Schick, 1994b). Before leaving this subject, we note that the form of the amphiphile–amphiphile structure function should be the same in the L_3 phase as in the microemulsion because the theories, and the physics, are the same (Strey *et al.*, 1991). A comparison of Fig. 1.11 and the data for $\psi = 0.0897$ in Fig. 3.17 makes this evident.

Perhaps the most interesting result from this theory that treats the difference between inside and outside water as an order parameter is that it leads to an inside/outside phase transition. This is clear because the two-component Landau theory is identical to that of the simple ternary mixture, or of the equivalent Blume–Emery–Griffiths model, and will therefore have the same phase transitions. The one major difference is that one has no field which couples to the difference between inside and outside waters, so that the system

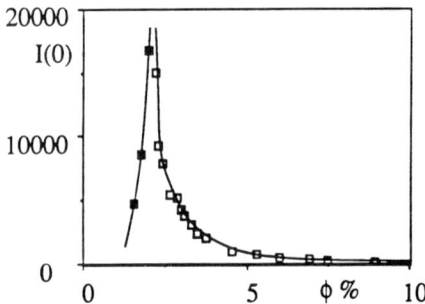

Fig. 3.20 Scattering intensity at zero wavevector on approaching the inside/outside transition. From Coulon *et al.* (1991).

is always in the symmetric subspace. When the order parameter is zero, the system is in the disordered state or symmetric sponge phase. The inside/outside symmetry can be spontaneously broken in which case there is much less inside water than outside water, or *vice versa*. This transition is the analogue of that to oil- and water-rich phases in the ternary mixture. In the binary mixture, the broken symmetry state can be thought of as representing a droplet phase. We know from the simple model of ternary mixtures that this transition can be continuous or first order, and that a tricritical point will separate these two behaviours. It seems clear that in the $C_{12}E_5$-water system shown earlier in Fig. 1.9 a first-order transition occurs. One sees from that figure that the L_3 phase shown coexists with an almost pure water phase. The former represents the symmetric sponge phase, the latter the asymmetric one. As the temperature is varied, these two phases do not become critical as seen by the fact that their amphiphile concentrations do not become equal. A continuous inside/outside transition has been observed in the system of ionic amphiphile SDS, pentanol, water and salt (Coulon *et al.*, 1991). The intensity of scattered light extrapolated to zero wavevector $I(0) \propto S(0)$ was obtained and is shown for one particular path in Fig. 3.20. From the form of the scattering given in (3.92), one sees that this intensity should diverge as the scattering length ξ diverges. It should be noted that in the symmetry broken phase in which $\phi \neq 0$, the calculation leading to (3.92) has to be modified, but this does not affect the form of the divergence (Roux *et al.*, 1992a). This transition is the most obvious sign that the intriguing analysis of the sponge phase in terms of an additional intrinsic order parameter, differentiating inside from outside, is correct. An apt analogy is made, *en passant*, by Roux *et al.* (1992a) between the isotropic sponge phase and liquid He^4 above its lambda point. Both fluids are characterized by an internal order parameter which is not coupled to an external field. A symmetry-breaking transition can occur resulting in a line of transitions which can be first or second order, with a tricritical point between them. Just as the superfluid transition is usually observed by thermodynamic measurements, so too could the inside/outside transition be observed in this way. An ordinary Ising specific heat signal is expected.

A valid concern about the inside/outside transition is the effect of holes in the bilayer, which have some non-zero thermal probability of occurring. Such holes render meaningless the distinction between inside- and outside-water regions and, therefore, the description employing the scalar order parameter ϕ. Their effect was considered by Huse and Leibler (1991) who made use of a local gauge theory to describe the fluctuating bilayer interface. They concluded that as long as the line tension is greater than some critical amount, the holes do not alter the existence or the universality class of the transition. Although the transition remains, the scalar order-parameter description does not, and it would be interesting to rederive the structure function within the lattice gauge description to see how the presence of such holes is manifested in the scattering.

3.4.2 Lyotropic phases

In order to describe the phase diagram of the binary system, one which includes several lyotropic phases, one must go beyond the Landau theory with two scalar order parameters employed above, and take into account more of the structure of the amphiphile; that is, one must use at least a vector order parameter to describe it. One way to do so is to take the binary limit of the three-order-parameter model of Section 3.2.3. In this limit, the two scalar order parameters play the same role, because an incompressible two-component system is described by a single concentration variable. Therefore one can set $\psi(\mathbf{r}) = \text{constant} - \phi(\mathbf{r})$ in (3.29). Then this functional reduces to

$$\mathcal{F}[\phi, \tau] = \int d^3r[\alpha_1(\nabla \cdot \tau)^2 + \alpha_2(\nabla^2\tau)^2 + \alpha_3(\nabla \times \tau)^2 + \alpha_4(\nabla(\nabla \cdot \tau))^2$$

$$+ \beta_1(\nabla\phi)^2 + \gamma(\tau \cdot \nabla\phi) + U(\phi, \tau^2) - \mu\phi]. \tag{3.95}$$

In order to account for the different molar volumes of water and amphiphile, Gompper and Klein (1992) chose

$$U(\phi, \tau^2) = f_1(\phi) + \tilde{B}_2\tau^2 + \phi\left[\tilde{C}_2\left(\frac{\tau}{\phi}\right)^2 + \tilde{C}_4\left(\frac{\tau}{\phi}\right)^4\right], \tag{3.96}$$

and

$$f_1(\phi) = \tilde{A}_2\phi^2 + T[\phi\ln\phi + M(1 - \phi)\ln(1 - \phi)]. \tag{3.97}$$

For $\tau = 0$, $U(\phi, \tau^2 = 0) = f_1(\phi)$ is identical with the usual Flory–Huggins form (de Gennes, 1979) of the free energy. The parameter T plays the role of a temperature in the model.

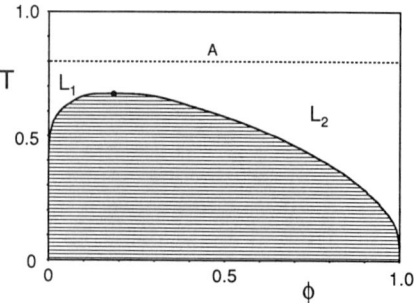

Fig. 3.21 Phase diagram from the functional of (3.95) to (3.97) with the parameters $\alpha_1 = -6$, $\alpha_2 = 9$, $\alpha_3 = 5$, $\alpha_4 = 5$, $\beta_1 = 10$, $\gamma = 10$, $M = 20$, $\tilde{A}_2 = -10$, $\tilde{B}_2 = 1$. The dot (•) marks the position of the critical point. From Gompper and Klein (1992).

When the amphiphiles are weak, i.e. when they do not form liquid crystalline phases, and have only a weak orientational interaction with water, the phase diagram looks typically as shown in Fig. 3.21: there is a lower miscibility gap between two disordered phases, a water-rich phase and an amphiphile-rich phase. The fact that the phases are disordered does not mean that they must be structureless. To obtain information about their structure, we consider again various scattering intensities. To do so, the free-energy functional is expanded to second order in the deviations of the order parameters from their average values $\tau = 0$ and $\phi = \phi_0$. This is most easily done in Fourier space, with the result

$$\mathcal{F} = \mathcal{F}_{MF}(\phi_0) + \int d^3k \begin{pmatrix} \tau_{\mathbf{k}} \\ \phi_{\mathbf{k}} \end{pmatrix} L(\mathbf{k}) \begin{pmatrix} \tau_{-k} \\ \phi_{-k} \end{pmatrix} + \dots \qquad (3.98)$$

where the matrix L is given by

$$L(\mathbf{k}) = \begin{pmatrix} K(\mathbf{k}) & -\frac{i}{2}\gamma\mathbf{k} \\ \frac{i}{2}\gamma\mathbf{k}^T & p^{-1}(\mathbf{k}) \end{pmatrix} \qquad (3.99)$$

with

$$K_{jl} = \zeta\delta_{jl} + \epsilon\frac{k_jk_l}{k^2} \qquad j,l \in \{1,2,3\}$$
$$p^{-1} = \beta_1 k^2 + \tfrac{1}{2}\partial_\phi^2 U(\phi,\tau^2)|_{\phi=\phi_0,\tau=0} \qquad (3.100)$$

and

$$\zeta = \alpha_3 k^2 + \alpha_2 k^4 + \tfrac{1}{2}\partial_\tau^2 U(\phi,\tau^2)|_{\phi=\phi_0,\tau=0}$$
$$\epsilon = (\alpha_1 - \alpha_3)k^2 + \alpha_4 k^4 \qquad (3.101)$$

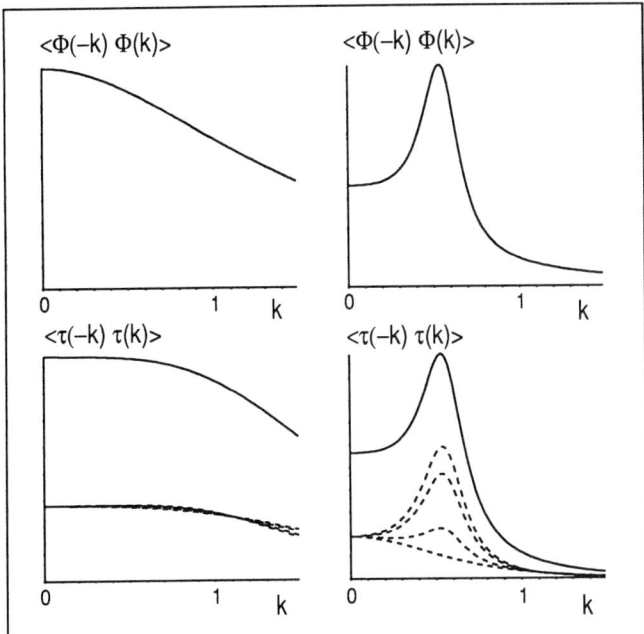

Fig. 3.22 Correlation functions in Fourier space at two-phase coexistence, for the temperature $T = 0.5$ in Fig. 3.21. The curves on the left show the correlations in the water-rich L_1-phase, the curves on the right in the amphiphile-rich L_2-phase. The dashed lines are the $\langle \tau_z(\mathbf{k})\tau_z(-\mathbf{k}) \rangle$ correlations for $\theta = 0$ (strongest peak at finite k), $\theta = \pi/6, \theta = 2\pi/6$ and $\theta = \pi/2$ (peak at $k = 0$). The parameters are the same as in Fig. 3.21. From Gompper and Klein (1992).

The inverse of the matrix L contains all structure functions. In particular, one finds (Gompper and Klein, 1992)

$$\langle \phi(\mathbf{k})\phi(-\mathbf{k}) \rangle = \frac{p(\zeta + \epsilon)}{\zeta + \epsilon - \frac{\gamma^2}{4}pk^2} \tag{3.102a}$$

$$\langle \tau(\mathbf{k}) \cdot \tau(-\mathbf{k}) \rangle = \frac{2}{\zeta} + \frac{1}{\zeta + \epsilon - \frac{\gamma^2}{4}pk^2} \tag{3.102b}$$

$$\langle \tau_z(\mathbf{k})\tau_z(-\mathbf{k}) \rangle = \frac{1}{\zeta + \epsilon - \frac{\gamma^2}{4}pk^2}\left[1 + \frac{\sin^2\theta}{\zeta}\left(\epsilon - \frac{\gamma^2}{4}pk^2\right)\right] \tag{3.102c}$$

where θ is the angle between $\mathbf{k}$ and the z-axis, i.e. $\cos^2\theta = k_z^2/k^2$.

The scattering intensities within the two phases at coexistence can now be compared. They are shown in Fig. 3.22 for $T = 0.5$. We see that the L_1-phase behaves like an ordinary fluid. The scattering intensity $\langle \phi(\mathbf{k})\phi(-\mathbf{k}) \rangle$ has a

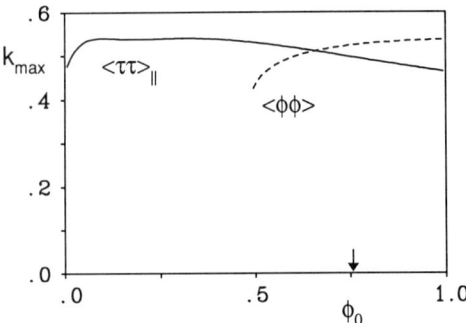

Fig. 3.23 Concentration dependence of the peak positions of the scattering intensities $\langle \tau_z(k_z)\tau_z(-k_z)\rangle$ (full line) and $\langle \phi(\mathbf{k})\phi(-\mathbf{k})\rangle$ (dashed line) along the path A in Fig. 3.21. From Gompper and Klein (1992).

peak at $k = 0$, and $\langle \tau_z(\mathbf{k})\tau_z(-k)\rangle$ is essentially independent of the angle θ. Thus in this phase, the amphiphiles do not form aggregates, and the fluid is disordered on a molecular scale. In the L_2-phase, on the other hand, the fluid is strongly structured, as indicated by the scattering peaks at non-zero wavevector of all correlation functions, except $\langle \tau_z(\mathbf{k})\tau_z(-k)\rangle$ for θ near $\pi/2$. This demonstrates that self-assembly has set in, although the phase is still homogeneous on a macroscopic scale.

The scattering behaviour *above* the critical temperature T_c is also interesting. The position of the scattering peak of the correlation functions $\langle \phi(\mathbf{k})\phi(-\mathbf{k})\rangle$ and $\langle \tau_z(k_z)\,\tau_z(-k_z)\rangle$ as a function of the amphiphile concentration is shown in Fig. 3.23. It can be seen that the amphiphiles are *always* oriented antiferromagnetically along their axis of orientation, leading to a peak at finite wavevector for all amphiphile concentrations. The peak position is essentially independent of that concentration. A peak at finite wavevector in the density–density correlation function, on the other hand, appears only at large amphiphile concentrations. It does not move out continuously from $k = 0$, but emerges from the shoulder of the scattering intensity at $k \simeq 0.4$. Furthermore, there is a range of amphiphile concentrations, where $\langle \phi(\mathbf{k})\phi(-\mathbf{k})\rangle$ has two peaks, one at $k = 0$, the other at finite k. This indicates a structure with three different length scales: the wavelength and correlation length of the bilayer structure, and the correlation length of the density fluctuations.

When the strength of the water–amphiphile and the amphiphile–amphiphile interaction is large enough, ordered phases can become stable. A typical phase diagram for the free-energy functional (3.95) to (3.97), is shown in Fig. 3.24. It contains four types of ordered phases. The simplest is the lamellar phase (L_α), a stack of bilayers of amphiphile separated by water layers, see

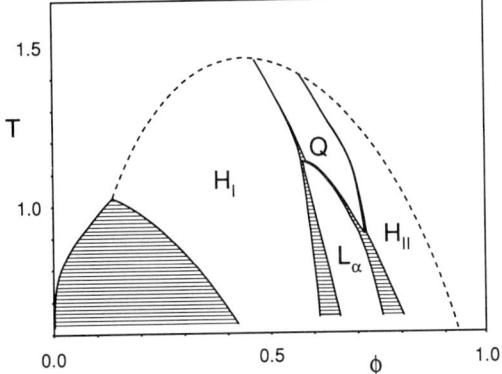

Fig. 3.24 Phase diagram in the temperature vs. amphiphile concentration plane for the free-energy functional (3.95) to (3.97) with the parameters $\alpha_1 = -6$, $\alpha_2 + \alpha_4 = 10$, $\beta_1 = 10$, $\gamma = 30$, $M = 20$, $\tilde{A}_2 = -10$, $\tilde{B}_2 = 1$, $\tilde{C}_2 = 17/12$, $\tilde{C}_4 = 1$. Phase transitions along dashed lines are second order. From Gompper and Klein (1992).

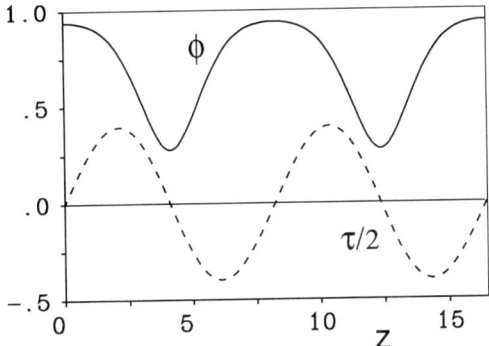

Fig. 3.25 Order-parameter profiles $\phi(z)$ and $\tau_z(z)$ for the lamellar phase (L_α) at $T = 0.75$, $\mu = -7.5$ of Fig. 3.24, with lattice constant $a = 8.235$ and average amphiphile concentration $\phi_0 = 0.703$. From Gompper and Klein (1992).

Fig. 3.25. The hexagonal phase (H_I) is a two-dimensional array of cylindrical micelles of infinite length. Its structure, as obtained from the model, is shown in Fig. 3.26. Similarly, the inverse hexagonal phase (H_{II}) is a two-dimensional array of water cylinders, separated by sheets of amphiphile, see Fig. 3.27. Finally, a stable simple cubic (sc) bicontinuous[†] phase is found, with a

---

[†]There is some ambiguity in the definition of a bicontinuous phase in this model. When constant concentration surfaces, with $\phi(\mathbf{r}) = \phi^*$ are used to divide space into water-rich and amphiphile-rich regions, the phase shown in Fig. 3.28 is bicontinuous for $0.58 < \phi^* < 0.80$.

Fig. 3.26 Contour plot of the amphiphile density distribution of the hexagonal phase (H_I) at $T = 0.75$, $\mu = -12.5$ of Fig. 3.24, with lattice constant $a = 9.386$ and average amphiphile concentration $\phi_0 = 0.562$. The interior of the cylinders is mainly amphiphile tails. From Gompper and Klein (1992).

structure shown in Fig. 3.28. Other cubic bicontinuous phases (Anderson *et al.*, 1990) may also be stable or metastable, but, due to the high numerical effort, have not been tested yet.

Thus, all types of ordered phases observed in experiments are present in the model. Moreover, with increasing amphiphilic concentration, one finds the sequence of phases hexagonal $\rightarrow$ lamellar $\rightarrow$ cubic $\rightarrow$ inverse hexagonal observed experimentally (Tiddy, 1980; Gruner, 1989; Seddon, 1990). For other values of the interaction parameters, some of these phases may be missing in the sequence.

Fig. 3.27 Contour plot of the amphiphile density distribution of the inverse hexagonal phase (H_{II}) at $T = 0.75$, $\mu = -5.0$ of Fig. 3.24, with lattice constant $a = 9.527$ and average amphiphile concentration $\phi_0 = 0.781$. The interior of the cylinders is mainly water. From Gompper and Klein (1992).

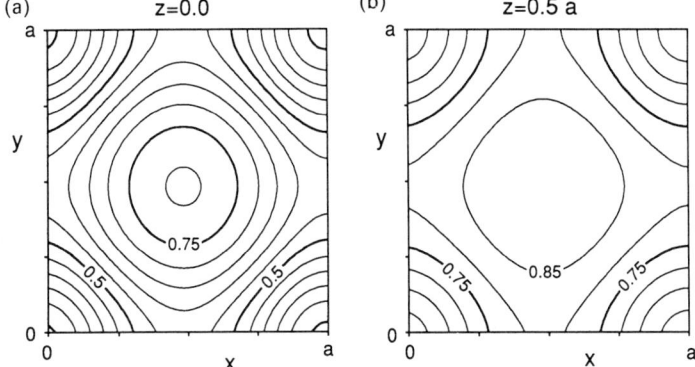

Fig. 3.28 The order-parameter distribution of the cubic (bicontinuous) phase (Q) at $T = 1.0$, $\mu = -9.0$ of Fig. 3.24, with lattice constant $a = 8.217$ and average amphiphile concentration $\phi_0 = 0.685$. (a) Contour plot of the amphiphile density distribution in the plane $z = 0$. (b) Contour plot of the amphiphile density distribution in the plane $z = a/2$. From Gompper and Klein (1992).

3.4.3   Elastic properties of bilayers of amphiphile

The same methods which have been developed in Section 3.3.3 for the calculation of the bending elasticity of amphiphilic monolayers at the oil/water interface can also be used to derive the bending elasticity of amphiphilic bilayers (Gompper and Klein, 1992). One again considers three different geometries: a planar bilayer, and a cylindrical and a spherical vesicle, and assumes that the real profile of the curved interfaces can be approximated, in the limit of weakly curved interfaces, by the planar profile. By calculating the Landau free energy of these order-parameter configurations as a function of the radius R, expanding in a power series in R^{-1}, and comparing with the corresponding expressions of the curvature Hamiltonian (3.52), one arrives at

$$\sigma = \int_{-\infty}^{\infty} \mathrm{d}z \, p_s(z) \tag{3.103a}$$

$$\lambda_s = 2 \int_{-\infty}^{\infty} \mathrm{d}z \, z \, p_s(z) \tag{3.103b}$$

$$\kappa = 2 \int_{-\infty}^{\infty} \mathrm{d}z \big[\alpha_1 \bar{\tau}_z^2 + 3(\alpha_2 + \alpha_4)[\bar{\tau}_z']^2 \big] \tag{3.103c}$$

$$\bar{\kappa} = \int_{-\infty}^{\infty} \mathrm{d}z \, z^2 \, p_s(z)$$

$$- 2 \int_{-\infty}^{\infty} \mathrm{d}r \big[\alpha_1 \bar{\tau}_z^2 + 4(\alpha_2 + \alpha_4)[\bar{\tau}_z']^2 \big] \tag{3.103d}$$

where

$$p_s(z) = 2\alpha_1[\bar{\tau}'_z]^2 + 4(\alpha_2 + \alpha_4)[\bar{\tau}''_z]^2 + 2\beta_1[\bar{\phi}']^2 + \gamma\bar{\tau}_z\bar{\phi}' \qquad (3.103e)$$

This result for the elastic moduli of amphiphilic bilayers shows strong similarities with the result for the amphiphilic monolayers in oil–water–amphiphile mixtures (Gompper and Zschocke, 1991). In particular, one sees that the surface tension σ and the spontaneous curvature modulus λ_s are the moments of the stress profile $p_s(z)$, as expected from general arguments on the elasticity of interfaces (Helfrich, 1981). However, there are important differences. First, the second moment of the stress profile cannot be written as a linear combination of κ and $\bar{\kappa}$. Second, since α_1 can be negative, it is not obvious that κ must always be positive in the present model. Finally it should be noted that the result (3.103c) for the bending rigidity depends mainly on orientation-dependent properties of the amphiphile (determined by the interaction constants α_1, α_2 and α_4), whereas the water–amphiphile interaction enters only via the form of the order-parameter profile.

3.5 Summary

We can summarize the results of this section once again in light of the three major topics that we raised in Section 1: phase behaviour, interfacial behaviour and structure of the microemulsion. As to the first, the Landau theory is successful in reproducing all the expected uniform phases as well as the non-uniform lyotropic phases. The phase diagrams are, like most of those produced by microscopic theories, in parameter spaces not directly accessible to experiment, and hence do not resemble those obtained in the laboratory. In principle, a closer connection could be made in the case of the one-order-parameter model, because in that case the parameters which enter the theory are directly related to the structure function. Hence one could, by means of this theory, correlate scattering data with all the quantities which can be derived from the Landau theory: phase diagrams, interfacial profiles, wetting behaviour, etc. The calculation of interfacial behaviour has to be counted as one of the successes of the Landau approach. These behaviours include the occurrence of oscillatory interfacial profiles, which seem to have been observed; the insight into the wetting behaviour which leads to a prediction of a first-order wetting transition whose location can be related to scattering data, and which has been experimentally confirmed; and the explanation of the behaviour of the contact angle as a function of temperature. Equally successful is the description of structure, as the results of the Monte Carlo simulations show. Such structure could in principle be obtainable from microscopic theories, but the simulation would have to be of an extremely large system. Because the length scale in the Landau–Ginzburg approach is not as small as in microscopic theories, the simulations can be carried out more easily. Finally, by providing the means to calculate elastic properties from underlying microscopic models, Landau–Ginzburg theories provide a bridge from such models to the larger length scale membrane theories, which are the subject of the next section.

4 The membrane approach

4.1 Introduction

When amphiphilic molecules self-assemble in a monolayer at the oil/water interface, they do so because the energy gain of the interactions exceeds the loss of entropy due to the assembly. As additional molecules are added to the system and go to the interface, the interfacial tension is reduced according to the Gibbs isotherm[†]

$$\Sigma^{-1} = -\frac{\partial \sigma(T, \mu)}{\partial \mu}, \tag{4.1}$$

where $\Sigma(T, \mu)$ is the area per molecule at the surface, or more strictly, the area per surface excess number of molecules. The interfacial tension is also related directly to the excess surface free energy per surface excess number of molecules, f_N, according to

$$\sigma = \frac{\partial f_N(T, \Sigma)}{\partial \Sigma}. \tag{4.2}$$

This reduction in tension continues as long as the molecules continue to go to the surface and decrease the area per molecule, Σ. Eventually this rapid decrease in the interfacial tension either changes significantly, or comes to an end altogether. The first occurs when the formation in the bulk of aggregates of amphiphiles, such as micelles, becomes more free-energetically favourable than an increase in the number of amphiphiles at the interface. With a small change in chemical potential, a large change in the concentration ϕ of amphiphiles can be accommodated in these aggregates, so that the decrease in tension with concentration

$$\frac{\partial \sigma(T, \phi)}{\partial \phi} = -\frac{1}{\Sigma} \frac{\partial \mu(T, \phi)}{\partial \phi}, \tag{4.3}$$

[†]For a review of surface thermodynamics, see Schick (1990).

becomes very small. The change in this derivative is one means of observing the critical micelle concentration. Even beyond this point, the interfacial tension should continue to decrease, although not as rapidly as formerly, because some of the added molecules do go to the surface. This decrease in tension absolutely ends, however, if a bulk phase transition to the micro-emulsion or to the lamellar phase is encountered, for at three-phase coex-istence, the temperature and chemical potential are fixed, which fixes $\sigma(T, \mu)$. Beyond this transition, the oil/water interface and its tension no longer exist. We have seen that this smallest value of interfacial tension which is reached can be made very low in systems with very strong amphiphiles. Even though it cannot vanish identically, because of the requirement of thermodynamic stability, it is useful to consider a fictitious reference state in which the tension does vanish. Presumably, such a state is not too far from the actual one for a very strong amphiphile. The interface of this fictitious state is called saturated, and one would have

$$0 = \frac{\partial f_N(T, \Sigma)}{\partial \Sigma}, \qquad \text{saturated interface.} \qquad (4.4)$$

(Unfortunately, this definition is often taken to be a general statement that the interfacial free energy is minimized with respect to the area per particle Σ, as if it were free to vary independently of the rest of the system. This is not correct, and (4.2) holds in general.) One can now ask what fluctuations contribute most importantly to the free energy of the saturated monolayer. The membrane approach, which is the subject of this section, takes the self-assembly of the interface as a given, constructs the free energy of the interface, and explores the consequences of that free energy. Because the self-assembly of the amphiphilic sheet is not an issue in this approach, the basic length scale will turn out to be much larger than molecular lengths. Hence the approach is well suited for examining large-scale configurations, such as vesicles, and fluctuations on very long wavelengths.

A similar argument can be applied to *internal* interfaces. If they are treated as two-dimensional objects, the analogues of (4.1) and (4.2) are

$$\Sigma^{-1} = \frac{\partial \pi(T, \mu)}{\partial \mu}, \qquad (4.5)$$

and

$$\pi = -\frac{\partial f_N(T, \Sigma)}{\partial \Sigma}, \qquad (4.6)$$

where π is the two-dimensional, or spreading, pressure. Note the difference in signs between π and σ. The two-dimensional pressure π must be positive. Just as the pressure of a self-bound three-dimensional liquid is very small, so the spreading pressure of a self-assembled internal interface can be very small.

This leads once again to the concept of a saturated system for which $\pi = 0$, to the question as to the dominant fluctuations, and to the membrane approach.[†]

4.2 Bending elasticity of fluid membranes

We begin by constructing the free energy for a two-dimensional fluid membrane which exists in a three-dimensional space.[††] The membrane's shape is described by a three-component vector field $\mathbf{X}(\nu)$, where $(\nu^1, \nu^2) = \nu$ are the internal coordinates of the membrane. It is assumed that the Hamiltonian, which describes the energy of deformations of the membrane, is a function of its shape only. In particular no internal degrees of freedom are necessary. This assumption is not true of course for a solid or crystalline membrane for which deformations from a preferred structure would have to be considered, just as they are in a three-dimensional solid (Nelson and Peliti, 1987; Aronowitz and Lubensky, 1988; Paczuski *et al.*, 1988). The free energy of the fluid membrane must be invariant under translations and rotations in three-dimensional space, and reparametrization invariant because the fluidity of the membrane ensures that no coordinate system is preferable to any other.

In order to understand the general form of the Hamiltonian, we need to recall some basic concepts of differential geometry. The tangent plane to a point P on the surface is determined by two tangent vectors whose covariant components are

$$\mathbf{t}_i = \partial_i \mathbf{X}, \qquad i = 1, 2 \tag{4.7}$$

with $\partial_i = \partial/\partial\nu^i$, and the ν^i are contravariant components of the vector ν. The two tangent vectors determine the metric tensor, or first fundamental form

$$g_{ij} \equiv \mathbf{t}_i \mathbf{t}_j = \partial_i \mathbf{X}\partial_j \mathbf{X}, \tag{4.8}$$

which relates the infinitesimal element of distance ds to the infinitesimal differences in coordinates,

$$\mathrm{d}s^2 = g_{ij}\mathrm{d}\nu^i\mathrm{d}\nu^j. \tag{4.9}$$

(We employ the Einstein convention in which summation over repeated indices is understood.) The inverse of the metric tensor g_{ij} is the tensor $g^{ij} = (g^{-1})_{ij}$. The metric tensor and its inverse relate covariant and contravariant components of any tensor, e.g. $T_{ij} = g_{ik}g_{jl}T^{kl}$. The infinitesimal

---

[†]For recent reviews see Nelson *et al.* (1989), Lipowsky (1991) and Lipowsky *et al.* (1992).
[††]See, for example, David (1989).

element of area, d^2S, generated by the two vectors $(d\nu^1, 0)$ and $(0, d\nu^2)$, is then

$$d^2S = \sqrt{\det g}\ d\nu^1 d\nu^2. \tag{4.10}$$

The generalization of a straight line on a surface is the geodesic; it is determined by the differential equation

$$\frac{d^2\nu^i}{ds^2} + \Gamma^i_{jk}(\nu)\frac{d\nu^j}{ds}\frac{d\nu^k}{ds} = 0, \tag{4.11}$$

where s is the distance along the curve. The quantity Γ^i_{jk}, the affine connection or Schwarz–Christoffel symbol, is given by

$$\Gamma^i_{jk}(\nu) = \frac{1}{2}g^{il}[\partial_k g_{jl} + \partial_j g_{lk} - \partial_l g_{jk}]; \tag{4.12}$$

it is not a tensor.

In order to define a derivative of a vector field in tangent space, we need the concept of parallel transport. Consider two points P and P' on the surface, and the geodesic $\nu(s)$ through these points (it is unique if the two points are close enough). A vector V in P is parallel transported along $\nu(s)$ to a vector V' in P', if (i) V and V' have the same length, and (ii) the angle between the vectors with the geodesic is the same. We can then define a "covariant derivative", D_j, of a vector field V by comparing the vector $V(\nu + \delta\nu)$ with the vector V' obtained by parallel transport from $V(\nu)$. This yields

$$D_j V^i \equiv \partial_j V^i + \Gamma^i_{jk} V^k. \tag{4.13}$$

The curvature of a surface in the embedding space is determined by the curvature tensor

$$\mathbf{K}_{ij} = D_i D_j \mathbf{X}. \tag{4.14}$$

Each component of the curvature tensor is a vector in bulk space. $\mathbf{K}_{ij}$ is proportional to the unit normal vector $\mathbf{n}$ at each point of the surface,

$$\mathbf{K}_{ij} = K_{ij}\mathbf{n}, \tag{4.15}$$

where K_{ij} is the second fundamental form of the surface. It determines the partial derivative of the normal vector (Weingarten equation),

$$\partial_i \mathbf{n} = K_{ij}\mathbf{t}^j. \tag{4.16}$$

At every point of the surface, K_{ij} may be diagonalized, and this defines the principal directions of curvature at this point. The eigenvalues are the inverse of the two principal radii of curvature, R_1 and R_2. Instead of the two principal radii of curvature, one often uses the following two scalar quantities, the

Gaussian curvature,

$$K = \det(K_{ij}) = \frac{1}{R_1 R_2}, \qquad (4.17)$$

and the mean curvature

$$H = \frac{1}{2}\text{Tr}(K_{ij}) = \frac{1}{2}\left(\frac{1}{R_1} + \frac{1}{R_2}\right). \qquad (4.18)$$

Unlike ordinary derivatives in flat space, covariant derivatives on a curved surface do not commute, and their commutator, $[D_i, D_j] \equiv D_i D_j - D_j D_i$, defines the Riemann curvature tensor

$$[D_i, D_j]V^k = R^k_{lij} V^l. \qquad (4.19)$$

This tensor has only a single independent component, the scalar curvature R, and can be written in the form

$$R^k_{lij} = \gamma^k_l \gamma_{ij} \frac{1}{2} R, \qquad (4.20)$$

with

$$\gamma_{ij} = \sqrt{\det g}\, \epsilon_{ij}, \qquad (4.21)$$

where ϵ_{ij} is the totally antisymmetric matrix. The Riemann curvature tensor has two remarkable properties:

(i) Gauss' *theorema egregium*: $R = 2K$
(ii) the Gauss–Bonnet theorem: the integral over a closed surface, S, is a topological invariant,

$$\int_S d^2\nu \sqrt{g} R = 4\pi\chi_E, \qquad (4.22)$$

where χ_E is the Euler characteristic, which counts the number of handles, g, of S according to $\chi_E = 2(1 - g)$.

To write down the elastic free energy as an expansion in derivatives of the vector field $\mathbf{X}(\nu)$, we consider the local invariants which can be formed from these derivatives. Up to second order in them, the invariants are 1, $\mathbf{n} \cdot \Delta\mathbf{X}$, $(\Delta\mathbf{X})^2$, and R, where the scalar Laplacian (Laplace–Beltrami operator), Δ, is given by

$$\Delta = D^i D_i. \qquad (4.23)$$

Integration of these local invariants over the entire surface area yields an expression which is invariant under translations and rotations, and is manifestly invariant under coordinate transformations. Hence we identify it with the elastic free energy of the surface. It is (Canham, 1970; Helfrich, 1973; Evans, 1974; Polyakov, 1986):

$$\mathcal{H} = \int d^2 \nu \sqrt{g} \left[\sigma + \frac{\lambda_s}{2} \mathbf{n} \cdot \Delta \mathbf{X} + \frac{\kappa}{2} (\Delta \mathbf{X})^2 + \frac{\bar{\kappa}}{2} R \right]. \tag{4.24}$$

From (4.14), (4.15), (4.18), $\mathbf{n} \cdot \Delta \mathbf{X} = 2H$, and $R = 2K$ from Gauss' *theorema egregium*, so that

$$\mathcal{H} = \int d^2 \nu \sqrt{g} \left[\sigma + \lambda_s H + 2\kappa H^2 + \bar{\kappa} K \right]. \tag{4.25}$$

This is a form which will play a major role in the remainder of this section, and is one which we have already used previously in Sections 3.3.3, and 3.4.3. Note that the coefficients σ, λ_s, κ and $\bar{\kappa}$ are defined as the coefficients in this expansion. In particular, there is as yet no thermodynamic meaning to the coefficient σ. If the Hamiltonian were describing an interface between two bulk phases, σ would be identified with an interfacial free energy. On the other hand, if internal interfaces were described, σ would be interpreted as the negative spreading pressure, $-\pi$.

4.3   Almost flat fluid membranes

Some important consequences of the elastic energy (4.25), can be seen even from the study of the relatively simple system of a nearly flat fluid membrane. We take the flat reference state to be parallel to the (x_1, x_2)-plane. Then, the membrane position can be described by a single-valued function $f(x_1, x_2)$, see Fig. 4.1. This is the Monge representation, with

$$\mathbf{X}(x_1, x_2) = (x_1, x_2, f(x_1, x_2)). \tag{4.26}$$

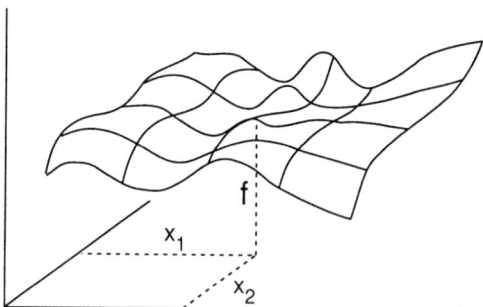

Fig. 4.1 The Monge parametrization of a surface is given by a single-valued function $f(x_1, x_2)$.

The unit normal vector at each point is easily shown to be

$$\mathbf{n}(x_1, x_2) = \frac{1}{\sqrt{g(x_1, x_2)}}(-\partial_1 f, -\partial_2 f, 1), \tag{4.27}$$

where

$$g(x_1, x_2) = 1 + (\nabla f)^2. \tag{4.28}$$

Since the tangent vectors at each point are given by

$$\mathbf{t}_1 = (1, 0, \partial_1 f), \qquad \mathbf{t}_2 = (0, 1, \partial_2 f), \tag{4.29}$$

one finds from (4.16)

$$\partial_1 n_1 + \partial_2 n_2 = \mathrm{Tr}(K_{ij}) = 2H. \tag{4.30}$$

Together with (4.27) and (4.28), this leads to

$$2H = -g^{-3/2} \big[\partial_1^2 f (1 + (\partial_2 f)^2) + \partial_2^2 f (1 + (\partial_1 f)^2) - 2\partial_1 f \partial_2 f \partial_1 \partial_2 f \big]. \tag{4.31}$$

We let the area of the membrane fluctuate without constraint by setting the coefficient $\sigma = 0$ in (4.25). Then, to leading order in an expansion in derivatives of f, the energy (4.25) reads (Brochard and Lennon, 1975)

$$\mathcal{H} = \frac{\kappa}{2} \int \mathrm{d}^2 x (\nabla^2 f)^2. \tag{4.32}$$

Following de Gennes and Taupin (1982), we can now estimate the fluctuations in the normals. The angle $\theta(x_1, x_2)$ which the normal $\mathbf{n}$ makes with the x_3 axis is given by

$$\cos \theta = \hat{\mathbf{x}}_3 \cdot \mathbf{n} = \frac{1}{\sqrt{g}}, \tag{4.33}$$

with $\hat{\mathbf{x}}_3^2 = 1$. Because (4.32) is a quadratic form, $\langle \theta^2 \rangle$ can be calculated by expanding both sides of (4.33) for small angles, and using the equipartition theorem. This is most easily done in Fourier space, and yields (de Gennes and Taupin, 1982)

$$\langle \theta^2 \rangle = k_B T \int \frac{\mathrm{d}^2 q}{(2\pi)^2} \frac{1}{\kappa q^2} \simeq \frac{k_B T}{\kappa} \ln(L/a), \tag{4.34}$$

where a is a microscopic length scale of the order of the molecular size, and L is the linear extension of the membrane. This result is the ubiquitous logarithmic divergence with system size which occurs in most two-dimensional systems with continuous symmetries (Nelson, 1989). Here it signals the breakdown of long-range order in the normals. Thus, even if one knows the orientation of the membrane at the origin of coordinates, one knows nothing about the orientation at distances far from the origin. One sees from this expression that the assumption of an almost flat membrane must fail at distances beyond some length at which the membrane has completely folded back in direction, i.e. $\langle \theta^2 \rangle \approx \pi^2$. Equation (4.34) says that this will happen at a distance of $a \exp(\pi^2 \kappa / k_B T)$.

Two theoretical approaches have been used to study the long wavelength behaviour of fluid membranes in more detail. One is a perturbation theory in the higher order derivatives in (4.31), which have been neglected in (4.32) (Peliti and Leibler, 1985; Förster, 1986; Kleinert, 1986b; David and Leibler, 1991). It can be viewed as a low-temperature expansion for fluid membranes, with the expansion parameter $k_B T / \kappa$. The other is an expansion in $1/d$, where d is the dimension of the embedding space (David, 1986; David and Guitter, 1987). This expansion is non-perturbative in the sense that it takes into account also far-from-flat configurations. However, both types of calculations neglect the effects of self-avoidance. The results of these calculations are first that the rigidity κ is renormalized by the fluctuations. At length scale L, to lowest order in perturbation theory, the renormalized rigidity $\kappa_R(L)$ is given by

$$\kappa_R(L) = \kappa - \frac{3k_B T}{4\pi} \ln \frac{L}{a} + \dots . \quad (4.35)$$

Thus the short-wavelength fluctuations reduce the rigidity at large length scales. Second, the normal–normal correlation function decays exponentially,

$$\langle \mathbf{n}(0) \cdot \mathbf{n}(\nu) \rangle \sim e^{-|\nu|/\xi_p}, \quad (4.36)$$

where ξ_p is the persistence length. It is obtained from (4.35) as the length at which κ_R vanishes, so that

$$\xi_p \simeq a \exp(4\pi\kappa/3k_B T). \quad (4.37)$$

This is essentially the same length that we estimated on the basis of (4.34). At length scales smaller than ξ_p, the membrane is effectively flat, i.e. its normals at different points are well-correlated. Beyond the length ξ_p, the normals are essentially uncorrelated, and the membrane is said to be crumpled. The fact that for any non-zero temperature the persistence length is always finite means that large enough fluid membranes *without self-avoidance* are always crumpled. We will see in Section 4.7.3 below what the crumpled state looks like, and also examine the effect of self-avoidance.

It should be noted that the fluctuating membrane is characterized by two areas, the real area A, and the area of its projection onto the (x_1, x_2)-plane, A_b (Golubović and Lubensky, 1989a; David and Leibler, 1991). These areas can either be held fixed, or be controlled by an external field. For almost flat membranes considered here, the simplest case is to keep the projected area fixed, and let the real area fluctuate by setting $\sigma = 0$. In this case one finds that the relative difference $(A - A_b)/A_b$ is also given by the integral (4.34), so that (Helfrich, 1985)

$$\frac{(A - A_b)}{A_b} = \frac{k_B T}{4\pi\kappa} \ln \frac{L}{a} + \dots . \tag{4.38}$$

4.4 Scaling considerations

Some interesting experimental conclusions can be drawn simply from the form of the elastic free energy (4.25), together with a few assumptions. We will assume that the spontaneous curvature modulus $\lambda_s = 0$, and for the moment ignore the first term. Thus we consider

$$\mathcal{H} = \int d^2 S [2\kappa H^2 + \bar{\kappa} K]. \tag{4.39}$$

An important point about this elastic energy is that it is scale invariant. That is, if all lengths are increased by a factor of b, then the area increases by b^2, but H^2 and K decrease by b^2 so that there is no effect on the energy. This means that the energy can only depend on the area of membrane A and the volume of the system V via the dimensionless combination A^3/V^2. This is rigorously correct. We now want to convert this to a statement about the concentration, ϕ, of amphiphile which is assumed to be the constituent of the membrane. Clearly ϕ is proportional to A/V, and this introduces a new length, l_0, the thickness of the membrane, $\phi = Al_0/V$. One does not know how this length should enter the elastic energy. No matter how it does so, it must invalidate the rigorous scaling argument. We ignore this difficulty and pursue the consequences which can be compared with experiment. Because ϕ is proportional to A/V, it scales as b^{-1}. (It is here that we are neglecting the length l_0.) The elastic energy is independent of b, so that the elastic energy per unit volume, f_{el}, scales as $1/V$ or as b^{-3}. With the assumption that the elastic energy dominates the free energy (per unit volume), f, the scaling implies (Porte *et al.*, 1991)

$$f = \phi^3 g(\kappa, \bar{\kappa}). \tag{4.40}$$

We can now dispense with the first term of the elastic energy (4.25). Because this term is linearly proportional to the area, it will by assumption be linearly proportional to ϕ, and can therefore simply be incorporated into the chemical potential of the amphiphile. There is one additional piece of information which should be included, and that is the effect of thermal fluctuations on the bending modulus κ. We saw in (4.35) that the renormalized bending modulus depends logarithmically on the size of the system L. A characteristic size of the membrane phase is $L = V/A = l_0/\phi$. Assuming that this provides a small correction to the leading term in (4.39), we can write for the free energy per unit volume

$$f = \phi^3[A + B\ln(\phi)]. \tag{4.41}$$

From this form of the free energy per unit volume, we can derive scaling laws for other quantities. For example, the osmotic pressure is given by (Porte *et al.*, 1991)

$$\Pi = \phi\frac{\partial f}{\partial \phi} - f,$$

$$= \phi^3(2A + B + 2B\ln\phi), \tag{4.42}$$

from which the intensity of scattered light at zero wavevector can be obtained:

$$I(0) \propto \phi\frac{\partial\phi}{\partial\Pi},$$

$$\propto [\phi(6A + 5B + 6B\ln\phi)]^{-1}. \tag{4.43}$$

Thus a plot of $[I(0)\phi]^{-1}$ vs. $\ln\phi$ should produce a straight line. Such a plot is shown in Fig. 4.2 with data taken from the sponge phase in the quasi-binary system AOT, water and salt. The agreement is good, but a greater variation in the concentration is needed to show conclusively the logarithmic form of the correction.

Similar arguments can be used to derive scaling laws for dynamic quantities such as the diffusion constant, D_c. Ignoring the renormalization of the

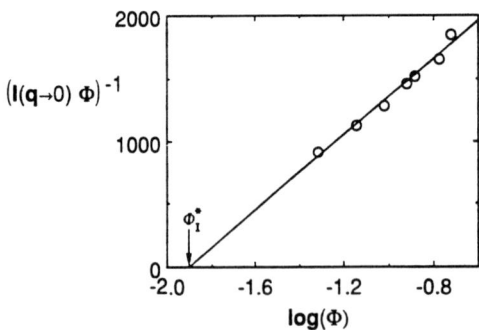

Fig. 4.2 A plot of $[I(0)\phi]^{-1}$ vs. $\ln\phi$ for the system AOT, water and salt. From Porte *et al.* (1991).

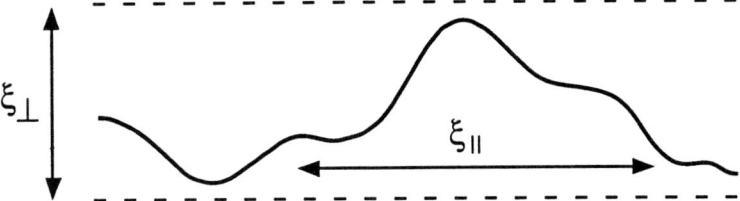

Fig. 4.3 Typical excursions of a fluctuating membrane are characterized by two length scales, $\xi_\perp$ and $\xi_\parallel$.

bending modulus, Porte *et al.* (1991) obtain $D_c \propto \phi$. This quantity was also measured in the AOT system, and found to obey this simple scaling without the need for renormalization effects. It is not clear why this should be so.

4.5   The Helfrich interaction

In addition to renormalizing the elastic moduli, the fluctuations of the membranes also bring about an entropic repulsion between them. In this section we derive this force. To do so, we consider a single membrane confined between two walls.

When a membrane is so confined, its transverse fluctuations are restricted which leads to an increase in the free energy. This increase can be estimated from the following argument. The transverse fluctuations of a finite piece of membrane, with diameter L, can be calculated for the quadratic Hamiltonian (4.32) from the equipartition theorem,

$$\xi_\perp^2 \equiv \langle f^2 \rangle_c = \langle f^2 \rangle - \langle f \rangle^2$$

$$= k_B T \int_{|q|>1/L} \frac{d^2 q}{(2\pi)^2} \frac{1}{\kappa q^4} = \frac{1}{4\pi} \frac{k_B T}{\kappa} L^2. \qquad (4.44)$$

Now, if the transverse fluctuations are restricted by the presence of two walls of separation $2d$, then $\sqrt{\langle f^2 \rangle_c}$ must be of the order of d; therefore, typical excursions in the membrane configurations, shown schematically in Fig. 4.3, will have a lateral extension of the size $\xi_\parallel$, with (Brochard and Lennon, 1975)

$$\xi_\parallel \sim \sqrt{\frac{\kappa}{k_B T}} \xi_\perp \sim \sqrt{\frac{\kappa}{k_B T}} d. \qquad (4.45)$$

This equation can be used to estimate the free energy (Lipowsky and Fisher, 1986; Lipowsky and Leibler, 1986, 1987). Since the typical excursions have the extensions $\xi_\perp$ and $\xi_\parallel$ perpendicular and parallel to the walls, respectively, the elastic bending energy per unit area must be of the order

$$H_{el} = \kappa(\nabla^2 f)^2 \sim \kappa(\xi_\perp/\xi_\parallel^2)^2. \tag{4.46}$$

There is also an entropic contribution due to the presence of the walls: every time the membrane hits the wall, it loses entropy of the order k_B (Fisher and Fisher, 1982). Therefore, the entropy loss per unit area is given by $S \sim -k_B/\xi_\parallel^2$. Thus, the free-energy difference between a confined membrane and a free membrane is given by

$$\Delta F = H_{el} - TS \sim \kappa(\xi_\perp/\xi_\parallel^2)^2 + k_B T/\xi_\parallel^2 \sim \frac{(k_B T)^2}{\kappa} d^{-2}. \tag{4.47}$$

This is a steric interaction between membranes due to their thermal undulations, one first calculated by Helfrich (1978). The same result can be obtained from an effective Gaussian Hamiltonian (Janke and Kleinert, 1986; Leibler and Lipowsky, 1987)

$$\mathcal{H}_G = \int d^2 x \left[\frac{\kappa}{2}(\nabla^2 f)^2 + \frac{\kappa}{2}\xi_\parallel^{-4} f^2 \right]. \tag{4.48}$$

where $\xi_\parallel$ now plays the role of a Lagrangian multiplier which is chosen in such a way as to make sure that $\xi_\perp \sim d$. Here, the perpendicular correlation length $\xi_\perp$ is given by

$$\xi_\perp^2 = k_B T \int \frac{d^2 q}{(2\pi)^2} \frac{1}{\kappa(q^4 + \xi_\parallel^{-4})} = \frac{1}{8} \frac{k_B T}{\kappa} \xi_\parallel^2. \tag{4.49}$$

In this model, the free-energy difference is

$$\Delta F = k_B T \int \frac{d^2 q}{(2\pi)^2} \ln\left(\frac{q^4 + \xi_\parallel^{-4}}{q^4} \right) = \frac{k_B T}{8} \xi_\parallel^{-2}. \tag{4.50}$$

With $\xi_\parallel$ related to $\xi_\perp$ according to (4.49), and $\xi_\perp^2 = \mu d^2$, with μ a constant of order unity,

$$\Delta F = c_1 \frac{(k_B T)^2}{\kappa} d^{-2}, \tag{4.51}$$

with $c_1 = 1/(64\mu)$. This has the same dependence on d as obtained in (4.47). Helfrich (1978) argues that $\mu = 1/6$ should be chosen (as the geometric mean between two simple extremal cases), so that $c_1 = 3/32 = 0.09375$. This result can be compared with that from Monte Carlo simulations (Gompper and Kroll, 1989; Janke et al., 1989; Janke, 1990) of a membrane between hard walls, described by the model of (4.32). By determining the change of the elastic energy with separation, one finds $c_1^{(MC)} = 0.0798 \pm 0.0005$, a somewhat smaller value. It is also possible to determine μ directly from the simulations by comparing $\xi_\perp$ with d (Gompper and Kroll, 1991). The Monte Carlo value is $\mu^{(MC)} = 0.130 \pm 0.002$, in the same range as Helfrich's estimate.

4.6 Phases and phase diagrams of membrane systems

4.6.1 General considerations of the phase diagram

On the basis of the elastic energy, (4.25), Huse and Leibler (1988) have proposed general types of phase diagram for a system with internal interfaces having no spontaneous curvature, $\lambda_s = 0$. In particular, they suggest for a system with $\bar{\kappa} = 0$ a schematic phase diagram as a function of σ and κ, as shown in Fig. 4.4, and for fixed $\kappa \gg k_B T$ a diagram in terms of σ and $\bar{\kappa}$, as shown in Fig. 4.5.

Let us first consider the phase diagram for $\bar{\kappa} = 0$. For large positive σ, the system does not want to have much interface present. It therefore breaks the symmetry between the two sides of the membrane, and forms two phases in equilibrium, each of which contains droplets of the minority component in a sea of the majority component. This is the "dilute droplet" phase. In the

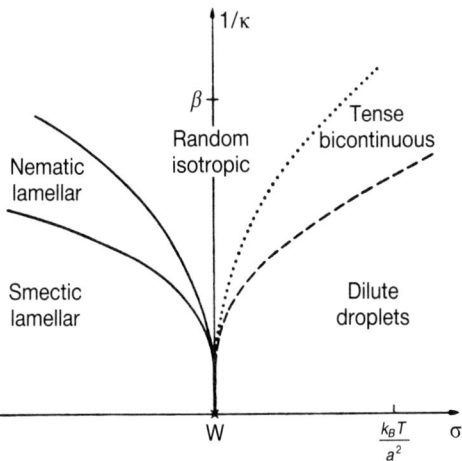

Fig. 4.4 Proposed schematic phase diagram of the membrane model for $\bar{\kappa} = 0$. All phase boundaries converge at the multicritical point at $\sigma = 1/\kappa = 0$. The boundary between tense bicontinuous and dilute droplet phases is a percolation transition rather than a thermodynamic one. The random isotropic phase is a fully disordered phase, while the ordering increases as one moves either towards the smectic lamellar phase or the dilute droplet phase. From Huse and Leibler (1988).

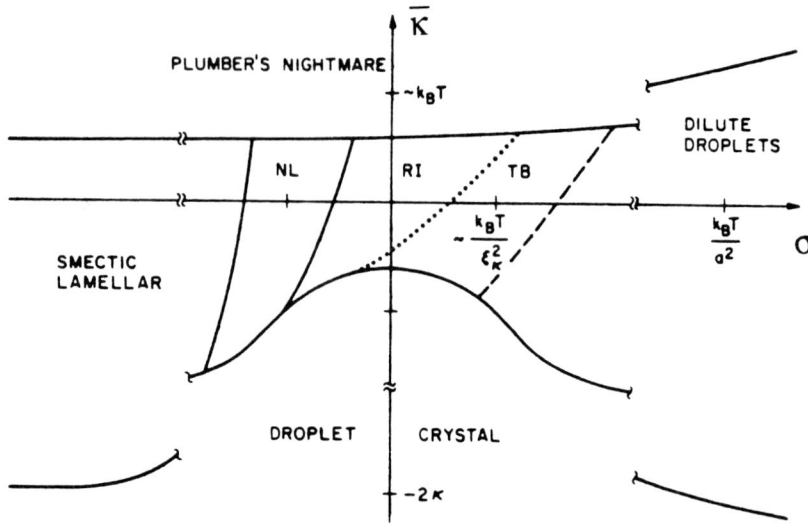

Fig. 4.5 Proposed schematic phase diagram of the membrane model as a function of σ and $\bar{\kappa}$, for $\kappa \gg k_B T$. Here, NL is a nematic lamellar phase, RI random bicontinuous and TB tense bicontinuous. From Huse and Leibler (1988).

ternary system, it would correspond to an oil- or water-rich phase, in the binary system to the asymmetric sponge phase. As σ decreases, the droplets of the minority component grow in size in order to increase the total film entropy, and form larger connected domains. These domains eventually span the system, forming what Huse and Leibler denote a "tense bicontinuous" phase. The transition to this phase, being a correlated percolation transition, shows no thermodynamic singularities. The oil–water symmetry is still spontaneously broken. As σ decreases still further, a transition to the disordered phase occurs in which the full symmetry of the system is restored. This transition can be first order, and if it is strongly so, the tense bicontinuous phase will not occur. This is almost certainly the case in balanced ternary systems with strong amphiphiles. The tense bicontinuous phase will certainly occur if the transition to the disordered phase is continuous, as in the inside/outside transition of the system SDS, pentanol, water and salt discussed earlier (Section 3.4.1). If the bending rigidity κ is large, and σ sufficiently negative, the system wants to put in as much membrane as possible, but not to bend it. A way to do this is to stack films parallel and close to one another. This produces the smectic lamellar phase with long-range orientational order and quasi-long-range order in the positions of the film planes. There is also the possibility that at intermediate negative values of σ, a nematic lamellar phase appears, with long-range orientational order, but only short-range positional order. These phases and their sequence is essentially what we have found in the microscopic and Landau–Ginzburg approaches. Here their progression is described in terms of different variables, the bending modulus κ and the interfacial tension σ, or spreading pressure $-\pi$.

We now take a system with a fixed bending rigidity $\kappa \gg k_B T$, and consider the possible phases as a function of the surface tension σ and the saddle-splay modulus $\bar{\kappa}$. For $\bar{\kappa} = 0$, we have of course the same sequence of phases as in Fig. 4.4. However, as $\bar{\kappa}$ is increased from zero, a new ordered phase is expected, one with vanishing mean curvature everywhere and negative Gaussian curvature.[†] Because of the proliferation of tubes and handles, the phase is referred to as the bicontinuous "plumber's nightmare"[††] (Scriven, 1976, 1977; Charvolin and Sadoc, 1987; Anderson et al., 1990; Charvolin, 1990). The membrane is a minimal surface in this case (Anderson et al., 1990). We have already encountered a similar phase in Section 3.4.2. When $\sigma = 0$ and $\bar{\kappa} > 0$, the simple Helfrich Hamiltonian (4.25) is unbounded from below, and is unstable with respect to a plumber's nightmare phase of infinitesimal lattice constant. For $\bar{\kappa}$ negative and of sufficiently large magnitude, it is energetically favourable to form many disconnected components, such as spherical droplets. When the droplets are very closed packed, they should freeze into a "droplet crystal". This behaviour was encountered in Section 3.3.5. For $\sigma = 0$ and $2\kappa + \bar{\kappa} < 0$, the Helfrich Hamiltonian is again unbounded, now with respect to a crystal of droplets of infinitesimal radius with an infinitesimal lattice constant.

If fluctuation effects are neglected, changes of conformation between spheres, cylinders and lamellae can be calculated quantitatively. This has been done by Safran et al. (1984), and by Wang and Safran (1990a). They consider a ternary system in the canonical ensemble, in which the volume fractions of oil, ϕ_o, and water, ϕ_w, as well as the volume fraction of amphiphile, ϕ, are controlled externally. The monolayer at the oil/water interface is characterized by a spontaneous curvature $c_0 = -\lambda_s/4\kappa$. The radii of monodisperse spheres and cylinders are determined by the ratio of the concentration of amphiphile to the concentration of the interior component, here assumed to be oil; $R_{sph}(w) = 3/(c_0 w)$ and $R_{cyl}(w) = 2/(c_0 w)$ with $w \equiv \phi/(\phi_o c_0 l_0)$ and l_0 the thickness of the amphiphile film. The free energies per unit area of lamellae, cylinders and spheres are, from (3.51), (3.52), (3.53)

[†]A positive value of $\bar{\kappa}$ also favours the formation of "passages" in the lamellar phase (Harbich et al., 1978; Huse and Leibler, 1988; Porte et al., 1989). Passages have indeed been observed experimentally (Harbich et al., 1978; Strey et al., 1992; Strey, 1992). The shapes and fluctuations of passages in lamellar phases have been calculated by Goos and Gompper (1993).
[††]Fluctuations of the plumber's nightmare phase have been studied by Bruinsma (1992).

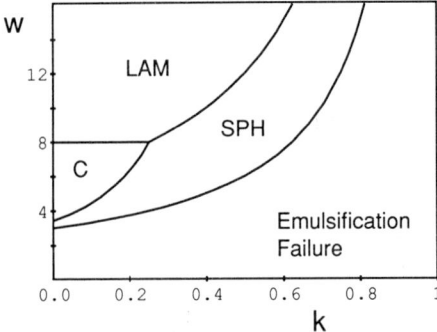

Fig. 4.6 The regions of stability of the spherical (SPH), cylindrical (C), and lamellar (LAM) phases of oil–water–amphiphile mixtures as a function of $w = (\phi/\phi_o)/(c_0 l_0)$ and $k = -\bar{\kappa}/2\kappa$.

with $\sigma = 0$,

$$f_{lam}/2\kappa c_0^2 = 1,$$

$$f_{cyl}/2\kappa c_0^2 = \left[\frac{w}{4} - 1\right]^2,$$ (4.52)

$$f_{sph}/2\kappa c_0^2 = \left[\frac{w}{3} - 1\right]^2 + \frac{1}{9}\frac{\bar{\kappa}}{2\kappa}w^2.$$

The region of parameter space in which a particular configuration has the lowest free energy per unit area is shown in Fig. 4.6. The origin of these conformational transitions is easy to understand. At low concentration of amphiphile, $w \ll 1$, the minimum free energy is that of spheres with $w_{sph} = 3(1 + \bar{\kappa}/2\kappa)^{-1}$. If the phase were uniform, w would be much less than this value, so the excess oil is rejected and coexists with a phase of monodisperse spheres[†] of radius $R_{sph}(w_{sph}) = c_0^{-1}(1 + \bar{\kappa}/2\kappa)$. The phase of excess oil is denoted "emulsification failure" in the figure. As the volume fraction of amphiphile is increased, the phase of monodisperse spheres is forced by the conservation conditions to accommodate the extra amphiphile, leading to a radius smaller than this preferred value and therefore to a higher free energy. At some point a phase of spheres with radius less than $c_0^{-1}(1 + \bar{\kappa}/2\kappa)$ has a higher free energy than a phase of infinite cylinders, which prefer the smaller radius $1/2c_0$. Thus the cylindrical phase becomes stable. With more amphiphile, the radius must again become smaller than this preferred value, and thus costly in bending energy. Eventually the lamellar phase, in which the bending energy is independent of w and thus of the amphiphile and oil concentrations, becomes the favourable one.

[†]The stability of the droplet phase when fluctuations are taken into account has been studied by Safran (1983, 1991).

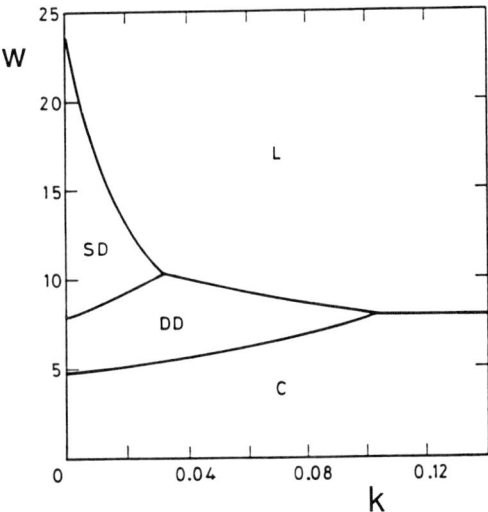

Fig. 4.7 The regions of stability of cylindrical (C), lamellar (L), single diamond (SD) and double diamond (DD) phase, as a function of $w = (\phi/\phi_o)/(c_0 l_0)$ and $k = -\bar{\kappa}/2\kappa$, for $\phi_o = 0.4$. From Wang and Safran (1990a).

The sequence of transitions determined by (4.52) depends only on the volume fraction ratio w, but not on the absolute volume fraction ϕ. However, the free energy of the ordered, bicontinuous structures of the single-diamond (SD) and double-diamond (DD) type does depend on ϕ explicitly (Wang and Safran, 1990a). This is because the periodicity of these structures enters as a fundamental length scale. Fig. 4.7 shows the phase diagram for a particular oil concentration in this case.

These phase diagrams are subject to the *caveat* that it is the free energy per unit volume which should be calculated, and the coexistence of phases determined from it in the usual way. This would entail, however, the inclusion of interactions between the membranes in order to stabilize the lamellar phase.

Of course these arguments can only be applied to well-characterized conformations, and cannot describe the disordered microemulsion. It is to the membrane description of this phase to which we now turn.

4.6.2   Phase diagram of the ternary system

There are several difficulties ecountered in obtaining a phase diagram of the ternary oil–water–amphiphile system based on the premise that the system consists of membranes governed by the elastic energy of (4.25). The first difficulty follows from the scaling behaviour. If all the amphiphile goes into making the membrane, then its concentration ϕ scales as the area per unit volume, that is, as the inverse of a characteristic length. As there are no characteristic lengths in the elastic energy, it follows that the elastic energy per unit volume depends on the concentration of amphiphile ϕ as

$$f_{el} = \phi^3\, g(\kappa, \bar{\kappa}). \tag{4.53}$$

In the absence of a microscopic length or of fluctuations, the entropy per unit volume in the random-mixing approximation also has this functional form. It immediately follows that, within mean-field theory, no phase transitions can occur simply by changing the concentration of amphiphile. This, of course, is patently contrary to experiment. Thus the effect of fluctuations, which vitiate this result, are not merely important in the membrane approach, they are essential. One could attempt to include these fluctuations via simulation, but this would be extremely difficult. Instead one puts in by hand the two most important effects of these fluctuations, the renormalization of the bending rigidity (4.35), and the Helfrich interaction, and treats the resulting effective free energy by mean-field theory.

Even with an effective free energy, it is still a difficult problem to consider a given configuration of membranes and calculate the energy and entropy of that configuration. To make this problem tractable, one makes use of a cell (Talmon and Prager, 1978) or lattice model (de Gennes and Taupin, 1982; Widom, 1984; Balbuena et al., 1986; Borzi et al., 1986; Safran et al., 1986). Approaches differ as to the properties of the amphiphile which are found on the surface of the cell, and whether the size of the cell is fixed or variable. We follow here the approach of Safran et al. (1986). Space is divided into cubic cells of lattice constant ξ, such that each cell is occupied either by oil or water, very much as in the lattice models[†] considered in Section 2. It is then assumed that whenever an oil-filled cell has a water-filled cell as its neighbour, the two are separated by a film of amphiphile. When it is assumed that the film at the oil/water interface forms an incompressible, two-dimensional liquid, the lattice constant ξ appropriate to a given phase is determined in terms of the concentration of amphiphile ϕ, and water ϕ_w of that phase. In the balanced system with zero spontaneous curvature, one has in the random-mixing approximation

$$\phi = \frac{6\phi_w(1 - \phi_w)l_0\xi^2}{\xi^3}, \tag{4.54}$$

so that

$$\xi = 6l_0\, \frac{\phi_w(1 - \phi_w)}{\phi}. \tag{4.55}$$

[†]A random surface model on a lattice with a discretized curvature Hamiltonian (4.25) has been studied by Mecke (1989), Gösmann (1990), Cappi et al. (1992) and Colangelo et al. (1993).

Here l_0 is again the thickness of the film. The free energy per unit volume of the whole system can then be written in a random-mixing approximation as a sum of two terms, the entropy of mixing, f_{mix}, and the bending energy of the film, f_b (Safran *et al.*, 1986; Andelman *et al.*, 1987; Cates *et al.*, 1988a). In the case of vanishing spontaneous curvature, it reads

$$f_{mix} = T\xi^{-3}[\phi_w \ln(\phi_w) + (1 - \phi_w)\ln(1 - \phi_w)] \qquad (4.56a)$$

$$f_b = \xi^{-3} 8\pi\kappa_R(\xi)\phi_w(1 - \phi_w). \qquad (4.56b)$$

The bending energy is obtained by associating the bends of the cubic lattice of size ξ with a spherical section of diameter ξ. Note that if the renormalized bending rigidity were independent of ξ, the free energy would be proportional to ϕ^3 due to the constraint (4.55), in agreement with the general scaling considerations. As noted earlier, if ϕ appeared nowhere else, there would be no phase transitions as the amphiphile concentration varied. The system would, for all concentrations, separate into oil- and water-rich regions. This is because in this model with a cell size which is not fixed at the outset, the entropy of mixing prefers a small cell size so that there is more entropy. From the constraint (4.55), this is accomplished in the pure water or oil phases (Widom, 1984). This result indicates that the membrane picture cannot be applied at small length scales. The amphiphile concentration does enter again through the fluctuations. At length scales smaller than ξ, fluctuations of the film are taken into account by using the renormalized bending rigidity κ_R (4.35), in (4.56b). This introduces the persistence length into the free energy, which is a function of two independent variables, ϕ_w and ϕ. Existence of uniform phases and coexistence between them now follows from standard manipulations. The lamellar phase is treated separately. It has neither entropy of mixing nor bending energy. Its energy is non-zero, however, due to the steric repulsion between the lamellae. Equation (4.51) is used for this energy, but with c_1 arbitrarily chosen to be 0.05. The phase diagram which is obtained is shown in Fig. 4.8. The phase diagram shows a single phase microemulsion region between a lamellar phase and a multi-phase region with both two- and three-phase equilibria. Along the two-phase boundary and in the middle phase, the characteristic length scale ξ of the microemulsion is of the order of the persistence length. The phase diagram is like that of a weak amphiphile rather than a strong one in that there exists in this balanced system a continuous path from the water-rich to the oil-rich phase. In the experimental phase diagram of a strong amphiphilic system, such as that shown in Fig. 1.7, the microemulsion is isolated. A comparison of the phase diagrams shows that it is most likely the treatment of the lamellar phase in the above calculation which is the source of the disagreement.

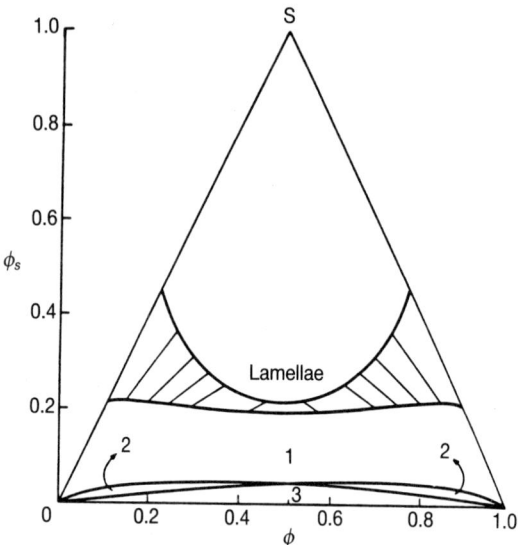

Fig. 4.8 Phase diagram of the membrane model for zero spontaneous curvature, at a temperature $3k_B T/(4\pi\kappa) = 0.2$. From Andelman *et al.* (1987).

The random-mixing approximation is not sufficient to calculate the scattering intensity of the microemulsion phase. Therefore, the bending energy has to be analysed more carefully for all possible configurations of four cells sharing one edge. This leads to a generalized Ising model with up to four-spin interactions, which can be treated within the mean-field and Ornstein–Zernike approximations (Milner *et al.*, 1988; Chandra and Safran, 1992). The result is a scattering intensity which has a peak at non-zero wavevector $q_{max} \simeq \pi/\xi$. Using (4.55), one obtains

$$q_{max} \simeq \frac{\pi}{6l_0} \frac{\phi}{\phi_w(1 - \phi_w)}. \tag{4.57}$$

Thus, the peak position moves out linearly with increasing amphiphile concentration. This seems to be consistent with experimental observations as seen, for example, in Fig. 1.11. Only a small concentration range is accessible in the experiments, however.

A unified treatment of the microemulsion and lamellar phases has been given by Golubović and Lubensky (1989b, 1990). The basic observation for their treatment is that in coarse graining the membrane fluctuations, the steric repulsion of membranes has to be taken into account. This can be done approximately by introducing hard walls parallel to the plaquettes of the cubic cell lattice, such that the fluctuations perpendicular to the plaquettes are restricted to $\pm\xi/2$. This leads to a coarse-grained Hamiltonian

$$H_{CG} = \int d^2 S \left[\sigma_{eff} + 2\kappa_{eff} H^2 + \bar{\kappa}_{eff} K \right], \tag{4.58}$$

where the radii of curvature in $H = \frac{1}{2}(R_1^{-1} + R_2^{-1})$ and $K = (R_1 R_2)^{-1}$ are now of the order of the lattice constant. For $R_1, R_2 \simeq \xi \gg \xi_{\parallel}$, where $\xi_{\parallel}$ (4.44), is the average distance between collisions of the membrane with the walls, the effective interfacial tension σ_{eff} and the effective bending rigidity κ_{eff} are given by

$$\sigma_{eff} = \sigma \left[1 + \frac{T}{4\pi\kappa} \ln(\xi_{\parallel}/a) \right] + \frac{\pi T}{4\xi_{\parallel}^2}, \tag{4.59}$$

$$\kappa_{eff} = \frac{3T}{4\pi} \ln(\xi/\xi_p). \tag{4.60}$$

The second term in (4.59) is the reduction in entropy due to the steric repulsion of membranes. The Gaussian curvature term is difficult to handle and is ignored; the saddle-splay modulus is set to zero. Finally, the coarse-grained Hamiltonian is written in terms of Ising cell variables $s_i = \pm 1$. To do this, an occupation number $N_{\langle ij \rangle}$ of broken bonds is introduced,

$$N_{\langle ij \rangle} = s_i(1 - s_j) + (1 - s_i)s_j, \tag{4.61}$$

which is 1 if the bond is broken, and 0 if it is not. The number of bends in a configuration of four cells $ijkl$ sharing a common edge is counted by the unique spin operator

$$N_{\langle ijkl \rangle} = N_{\langle il \rangle}N_{\langle ji \rangle} + N_{\langle ji \rangle}N_{\langle kj \rangle} + N_{\langle kj \rangle}N_{\langle lk \rangle}$$
$$+ N_{\langle lk \rangle}N_{\langle il \rangle} - 2N_{\langle il \rangle}N_{\langle ji \rangle}N_{\langle kj \rangle}N_{\langle lk \rangle}, \tag{4.62}$$

where ij, jk, kl, and li are nearest neighbours (and ik and jl are next-nearest neighbours). This operator can take the values 0, with four cells of the same type or a lamellar configuration with two neighbours of the same type, 1, with one cell different from the other three, or 2, with two cells of each type but as next-nearest neighbours. Thus one arrives at the generalized Ising model

$$H_I = \sigma_{eff} A(\{s_i\}) + \kappa_{eff} B(\{s_i\}), \tag{4.63}$$

where

$$A(\{s_i\}) = \xi^2 \sum_{\langle ij \rangle} N_{\langle ij \rangle}$$

$$B(\{s_i\}) = \frac{\pi}{3} m \sum_{\langle ijkl \rangle} N_{\langle ijkl \rangle}. \tag{4.64}$$

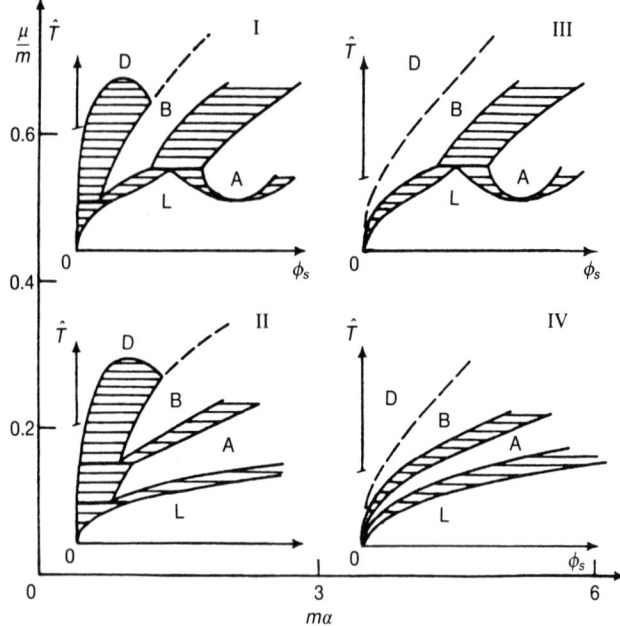

Fig. 4.9 Schematic representation of four types of $(\phi, k_B T/\kappa)$ phase diagrams calculated from model (4.59)–(4.64). The typical scale for ϕ is of order 0.1, while that of $k_B T/\kappa$ is of order 1. The region of coexisting oil- and water-rich phases is denoted D, the uniform disordered phase by B, lamellar by L, and a cubic, ordered phase by A. Cross-hatched regions show two-phase coexistence, while dashed lines indicate second-order transitions. The parameter α is a prefactor in (4.35). It is chosen to take the value 3, which agrees with that equation. From Golubović and Lubensky (1989b).

Here, a phenomenological parameter m is introduced to account for some ambiguities in assigning a smooth interface configuration to a given distribution of Ising spins. The value $m = 2$ was used by Milner *et al.* (1988), but other values could suffice.

The model (4.59)–(4.64) is now studied within the usual mean-field approximation. Equation (4.49) is used to express $\xi_{\parallel}$ in terms of $\xi_{\perp}$, and the latter is written $\xi_{\perp}^2 = \mu d^2$ with $d \equiv \xi/2$. The theory then depends on two phenomenological parameters, μ and m. Since the values for these parameters are uncertain, phase diagrams have been determined for various values of μ and m. Four types of phase diagrams are then obtained, as shown in Fig. 4.9. Each is shown in the temperature, amphiphile concentration space. The dimensionless temperature $\hat{T} \equiv k_B T/\kappa$. All these phase diagrams show the following features:

(1) The lamellar phase dominates at low temperatures, whereas other ordered phases can occur at higher temperatures.

(2) The transition between oil- and water-rich phases and the disordered phase is continuous at high temperatures and, for small $m\alpha$, becomes first-order at low temperatures. The two regimes are separated by a critical endpoint.

(3) The smallest amount of amphiphile is needed to solubilize oil and water near the four-phase point of oil-rich and water-rich phases, disordered, and lamellar phase.

The type-I diagram best reproduces the behaviour observed in experiment: a first-order transition from oil–water coexistence to the microemulsion phase, followed by a first-order transition to the lamellar phase with increasing amphiphile concentration. However, instead of the observed tricritical point followed by three-phase coexistence, the calculated diagram has a critical endpoint and a region of coexistence of four homogeneous phases: two different oil-rich phases, and two different water-rich phases. This has never been observed. Aside from this difference, this phase diagram in the $(T/\kappa, \phi)$-plane resembles that calculated from the three-component model and shown in the (amphiphile strength L, ϕ)-plane in Fig. 2.4. Because no phase diagrams have been calculated in the full concentration space for a fixed temperature, one cannot determine if the unified treatment of the homogeneous and lamellar phases has brought about a significant improvement in the location of the lamellar phase and has caused the disordered phase to be isolated, as observed in experiment.

4.6.3   The lamellar phase

The lamellar phase is particularly interesting because the effect of fluctuations is such that there is no long-range positional order, only an orientational order of the lamallae; i.e. the phase is completely analogous to a smectic phase of liquid crystals. To see this, we describe the system as follows. On length scales large compared with the average separation of two membranes, the fluctuations can be described by a coarse-grained displacement variable $u(x_1, x_2, z)$, where (x_1, x_2) are the coordinates parallel to the membranes, and z is the perpendicular coordinate. The system can then be described by an effective continuum model (Helfrich, 1978; Leibler and Lipowsky, 1987), with the Landau–Peierls Hamiltonian

$$\mathcal{H}_{LP} = \int d^2xdz \left[\frac{1}{2}B\left(\frac{\partial u}{\partial z}\right)^2 + \frac{1}{2}K(\nabla^2 u)^2 \right], \qquad (4.65)$$

where $K = \kappa/d$ is the bending rigidity of the whole stack, and

$$B = d \frac{\partial^2}{\partial d^2} \Delta F \tag{4.66}$$

is the vertical compressibility. This model has been introduced first in the context of smectic liquid crystals (see, for example, de Gennes (1974)). That there can be no true long-range order in the system can be seen most easily by calculating the variance $\langle u^2 \rangle$ of a single membrane in a stack of finite thickness, D. The variance is given by (Janke and Kleinert, 1987)

$$\langle u^2 \rangle = k_B T \int_{\pi/D}^{\pi/a} \frac{dq_z}{2\pi} \int \frac{d^2 q_\parallel}{(2\pi)^2} \frac{1}{B q_z^2 + K q_\parallel^4} = \frac{k_B T}{8\pi} \frac{1}{\sqrt{BK}} \ln(D/a), \tag{4.67}$$

where a is some short-distance cutoff (such as the thickness of a membrane). This logarithmic divergence indicates the loss of translational order in the limit $D \to \infty$.

This quasi-long-range order can be observed by analysing the intensity peaks from scattering measurements. In the vicinity of an mth order Bragg-position, $q_m = 2\pi m/d$, the scattering is given by (Caillé, 1972)

$$\begin{aligned} S(q_z, q_\parallel = 0) &\sim (q_z - q_m)^{-(2-X_m)}, \\ S(q_z = 0, q_\parallel) &\sim q_\parallel^{-(4-2X_m)}, \end{aligned} \tag{4.68}$$

with an exponent

$$X_m = q_m^2 \frac{k_B T}{8\pi\sqrt{BK}}. \tag{4.69}$$

Within a lamellar phase, the free energy ΔF of a single membrane has the same form as in (4.51), but with a new coefficient c_∞, which takes into account that there is a difference in confinement between hard walls and between fluctuating membranes. Using this expression in (4.66) and (4.69) and the definition of q_m, we obtain

$$X_m = m^2 \frac{\pi}{2\sqrt{6c_\infty}}. \tag{4.70}$$

This is a remarkable result, because it shows that the scattering exponent is independent of the membrane separation d, and can therefore be expected to have a universal value. In fact, when the thickness, δ, of the membranes is taken into account, a correction term appears, which gives (Helfrich, 1978)

$$X_m(\delta) = X_m(1 - \delta/d)^2. \tag{4.71}$$

Thus the scattering exponent approaches its universal value as $\delta/d \to 0$. Strictly speaking, the result (4.70) is the leading term in a low temperature expansion. Corrections to first order in the temperature have been calculated (Golubović and Lubensky, 1989a), and X_m is found to increase.

To evaluate the exponent, one needs the value of c_∞. Helfrich (1978) uses the expression (4.65) to calculate the free energy of the lamellae as a function of K and B. He finds that (4.66) for B is self-consistent if

$$c_\infty = 3\pi^2/128 = 0.2313... \qquad (4.72)$$

which implies a scattering exponent

$$X_1 = 4/3. \qquad (4.73)$$

The universal constant c_∞ can also be estimated by considering a single membrane between two walls, and gradually increasing the number of membranes such that the membrane separation stays constant. The analogy with domain-walls in $(1 + 1)$ dimensions leads to the extrapolation hypothesis (Gompper and Kroll, 1989)

$$c_n = \frac{2(2n + 1)}{3(n + 1)} c_1. \qquad (4.74)$$

This result is in good agreement with Monte Carlo simulations for three and five membranes, which give $c_3 = 0.0931$ and $c_5 = 0.0975$ (Gompper and Kroll, 1989). Therefore, we can extrapolate with some confidence to

$$c_\infty^{(MC)} = 0.106, \qquad (4.75)$$

which implies

$$X_1^{(MC)} = 1.97. \qquad (4.76)$$

The same result has been obtained by Janke et al. (1989). A third estimate has been obtained from a functional renormalization group study of an infinite stack (David, 1990). In this case one finds

$$c_\infty^{(RG)} = 0.081 \pm 0.025, \qquad (4.77)$$

so that

$$X_1^{(RG)} = 2.25 \pm 0.30, \qquad (4.78)$$

which is in agreement with the Monte Carlo estimate. Both the MC and the RG values for X_1 are considerably larger than Helfrich's estimate (4.73).

Lamellar phases of fluid membranes have been studied intensively by X-ray scattering in the last few years. After the pioneering work of Safinya et al. (1986) and Larche et al. (1986), several studies have shown that the scattering intensity is indeed characterized by algebraic singularities of the form (4.68), with an exponent X_1 in agreement with (4.71) (Bassereau et al., 1987, 1991;

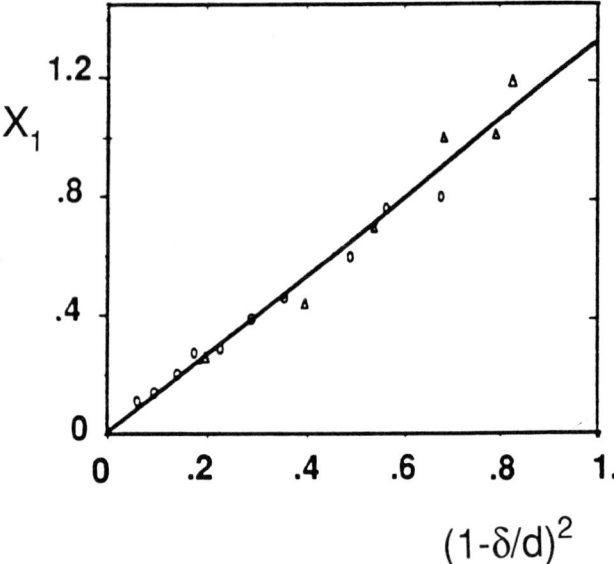

$$(1-\delta/d)^2$$

Fig. 4.10 The exponent X_m which characterizes the power-law behaviour of the scattering intensity near the Bragg-positions, as a function of the scaled distance d between the membranes. The thickness of the membrane is δ. From Roux and Safinya (1988).

Porte *et al.*, 1988a; Nallet *et al.*, 1990). This is shown in Fig. 4.10. It has also been shown that this behaviour is due to the undulation modes of the membranes, since in the case of electrostatic repulsion between membranes, which is longer ranged than the steric repulsion, the membranes become completely flat, and the scattering behaviour is no longer given by (4.68) (Roux and Safinya, 1988; Nallet *et al.*, 1990). Furthermore, the exponent X_m has been measured for several different systems, and was found to be universal (Safinya *et al.*, 1989). The value of X_1 obtained from almost all these experiments is

$$X_1^{(exp)} = 4/3. \qquad (4.79)$$

This is exactly the value (4.73) obtained by Helfrich's self-consistency argument. At present, it is unclear why it differs so much from the presumably more reliable estimates (4.76) and (4.78) obtained from Monte Carlo simulations and renormalization group theory.

X-ray and neutron scattering have also been used to measure the repeat distance, d, of dilute lamellar phases as a function of the amphiphile volume fraction ϕ (Strey *et al.*, 1990; Roux *et al.*, 1992b). In systems dominated by undulation forces, a deviation is found from the simple relation $\phi \propto d^{-1}$

which assumes flat interfaces. This deviation is in agreement with the logarithmic difference between the actual and projected areas of (4.38).

Stacks of fluid membranes have also been studied using the surface force apparatus (Israelachvili and Adams, 1978). The measurements (Kékicheff and Christenson, 1989; Richetti *et al.*, 1990; Abillon and Perez, 1990) confirm the prediction (4.51) of a d^{-2} behaviour of the free energy for systems dominated by fluctuations, and give a d^{-3} behaviour for systems with electrostatic repulsion. In the former case, the amplitude is again found to be consistent with Helfrich's estimate (4.72) (Abillon and Perez, 1990).

4.7 Vesicles

We have argued before that the hydrophobic effect is responsible for the formation of amphiphilic bilayers in water; this minimizes the contact of the hydrocarbon tails with water. However, for a flat piece of membrane, the tails are still exposed to water contact at the membrane's edge. This leads to an effective line tension which, for large enough membranes, pulls the edge together and closes the membrane (Helfrich, 1974; Litster, 1975; Drouffe *et al.*, 1991; Boal and Rao, 1992b). The result is a vesicle, a closed surface whose topology is usually that of a sphere.

4.7.1 Experiments on the elastic moduli of membranes

Several different methods have been used to determine the bending rigidity of the fluid membrane which comprises the vesicle. One method is to suck the vesicle part way into a micropipette (Evans and Rawicz, 1990). This increases the pressure and thus the lateral tension of the membrane. A measurement of the projected area as a function of the pressure yields the bending rigidity κ. Another method is to analyse the undulation spectrum of vesicles in terms of spherical harmonics (Schneider *et al.*, 1984; Duwe *et al.*, 1987; Bivas *et al.*, 1987; Milner and Safran, 1987; Faucon *et al.*, 1989; Duwe *et al.*, 1989, 1990; Méléard *et al.*, 1992; Zilker *et al.*, 1992). In this case, the bending rigidity can be determined from the mean amplitudes of the eigenfunctions. A similar analysis is also possible for flat interfaces (Meunier, 1985; Binks *et al.*, 1991; Lee *et al.*, 1991) and membranes (Mutz and Helfrich, 1989). The undulation spectrum in the lamellar phase can be measured by quasi-elastic light scattering (di Meglio *et al.*, 1985), by electron spin resonance (di Meglio and Bassereau, 1991), or by nuclear magnetic resonance (Bloom and Evans, 1991; Stohrer *et al.*, 1991). The bending rigidity can also be measured by deforming a vesicle in an electric field (Kummrow and Helfrich, 1991). In this case the shape of a vesicle in an electric field has to be calculated, and then compared with the experimental shape.

The result of all these experiments is that the bending rigidity of bilayers of biologically relevant phospholipids are of the order $\kappa \simeq 10k_B T$. Thus, biological membranes are very stiff, with a persistence length much larger than any other typical length scale. The shape of vesicles with these membranes is therefore given by the minimum of the elastic free energy of (4.25). Membranes with much smaller bending rigidity can be obtained by adding a small amount of bola-lipid to the membrane. In this case, bending rigidities of the order $\kappa \simeq k_B T$ have been observed (Duwe et al., 1990). Low bending rigidities have also been observed for ternary mixtures forming microemulsions, like $C_{12}E_5$, $C_{12}E_6$, AOT and SDS (di Meglio et al., 1985; di Meglio and Bassereau, 1991; Binks et al., 1991).

The measurement of the saddle-splay modulus is also possible, although more indirect (Farago et al., 1990; Meunier and Lee, 1991; Kellay et al., 1993).

4.7.2   The shape of stiff fluid vesicles

When the bending rigidity is much larger than $k_B T$, the persistence length is much larger than the diameter of a vesicle. In this case thermal fluctuations can be ignored, and the shape of a vesicle is determined by the minimum of the elastic energy. However, this does not imply that all vesicles are perfect spheres—there are important constraints, which have to be taken into account. The area A of the vesicle is fixed because, for all biologically relevant lipids, there is very little exchange of molecules between the bilayer and the surrounding fluid on typical experimental time scales. The volume V is also fixed because the volume of water enclosed changes negligibly on typical experimental time scales. Finally in some cases, the exchange of lipids between the inner and outer monolayers of the membrane is slow. In this case, the area difference, ΔA, between inside and outside membranes is fixed.

With these constraints, the minimization of the curvature energy produces a large variety of different vesicle shapes (Deuling and Helfrich, 1976; Peterson, 1985a,b; Svetina and Zeks, 1989; Berndl et al., 1990; Seifert et al., 1991; Miao et al., 1991). A few typical shapes are shown in Fig. 4.11, all of them axisymmetric by construction. Since the bending energy is scale invariant, the shape is determined by two dimensionless parameters, $v = (3V/4\pi)(A/4\pi)^{-3/2}$ and $\Delta a = (\Delta A/8\pi l_0)(A/4\pi)^{-1/2}$, where l_0 is the distance between the inner and the outer monolayer. The vesicle shown in Fig. 4.11(c) has the characteristic shape of a red blood cell (Elgsaeter et al., 1986). The observed variation of vesicle shapes with temperature (which changes v and Δa) has been reproduced by the curvature model (Berndl et al., 1990).

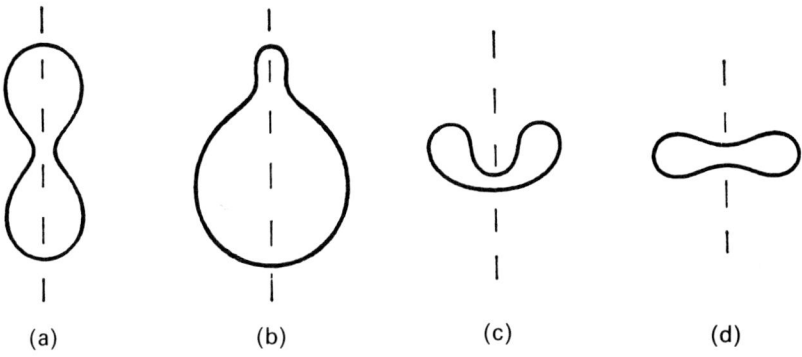

(a)   (b)   (c)   (d)

Fig. 4.11 The characteristic shape of vesicles as determined by the bending elasticity of the membrane, together with constraints on area, volume and the area difference between inner and outer monolayers. (a) Dumb-bell-shaped; (b) pear-shaped; (c) stomatocyte; (d) discocyte. The shapes are axisymmetric with respect to the dashed line. From Berndl *et al.* (1990).

Large vesicles usually do not form spontaneously and, when formed, decay with time. However, under certain conditions, large vesicles can be thermodynamically stable. This has been observed (Kaler *et al.*, 1989) for membranes made from mixtures of two ionic amphiphiles with oppositely charged head groups. This was later explained with a simple model (Safran *et al.*, 1990, 1991), in which the two amphiphiles have different spontaneous radii of curvature. It is found that vesicles form in which the concentration of the two components in the inner and outer monolayer of the membrane are different.

4.7.3   Monte Carlo simulations of floppy fluid vesicles

We have seen in Section 4.3 that on length scales larger than the persistence length, the membrane has no longer a well-defined orientation. If the bending modulus is very small, then so is the persistence length, and fluctuations of the vesicle membrane become important. Large fluctuations are particularly difficult to deal with analytically as any two parts of the membrane will experience a strong repulsion (self-avoidance) on contact with one another.

The model for the simulation of fluid membranes begins with a simple tether-and-bead model (Kantor and Nelson, 1987a,b; Kantor *et al.*, 1986, 1987; Plischke and Boal, 1988). It consists of a triangular network of N spherical beads of diameter $\sigma = 1$. Neighbouring beads in the network are linked by tethers of length l_0. Self-avoidance is generated by the pairwise hard-core repulsion of all beads, together with a choice of tether lengths $l_0 < \sqrt{3}$ and a sufficiently small stepsize s for each trial move. The number of

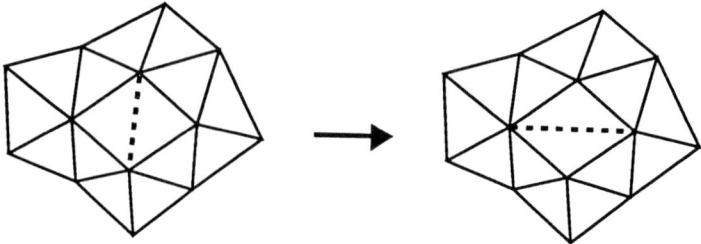

Fig. 4.12 The elementary Monte Carlo step which makes a tether-and-bead model a model for fluid membranes: a tether is cut, and reattached between the beads of the two adjacent triangles.

beads in typical simulations ranges from $N = 50$ to $N = 500$. The bending elasticity of the membrane is usually modelled by the product of surface normals of neighbouring triangles, so that

$$\mathcal{H}_{bend} = -\kappa \sum_{\langle i,j \rangle} (\mathbf{n}_i \cdot \mathbf{n}_j - 1) \tag{4.80}$$

Here, the sum is over all pairs of neighbouring triangles. What has to be added to this model is the fluidity of the membrane, i.e. the neighbour relations themselves have to become dynamical degrees of freedom, in addition to the vertex positions. Such models have been first used in the context of high-energy physics, where fluctuating surfaces appear in string theories (Polyakov, 1986, 1988). The basic idea is to include a new type of Monte Carlo step, in which a tether is cut and reattached as shown schematically in Fig. 4.12 (Ambjørn *et al.*, 1985, 1986; Boulatov *et al.*, 1986; Billoire and David, 1986). The surfaces in string theory are not self-avoiding. Hard spheres of finite radius together with a finite (and small enough) tether length are used to model self-avoiding surfaces (Ho and Baumgärtner, 1990; Baumgärtner and Ho, 1990). In addition to the bending rigidity x, the pressure difference $\Delta p = p_{in} - p_{out}$ is controlled in the simulations.[†]

The simulations for $\kappa = \Delta p = 0$ show very clearly that the vesicle is in a phase, which is characterized by tree-like, ramified conformations, as shown in Fig. 4.13. A scaling analysis (Kroll and Gompper, 1992a,b; Boal and Rao, 1992a) shows that this phase is in the *self-avoiding branched polymer* universality class, with the scaling behavior of volume V and radius of gyration R_g

$$\langle R_g^2 \rangle \sim N^{\nu_{bp}}$$
$$\langle V \rangle \sim N^{\nu_{bp}} \tag{4.81}$$

[†]For a brief review of recent results for floppy fluid vesicles, see Kroll and Gompper (1993).

pressure

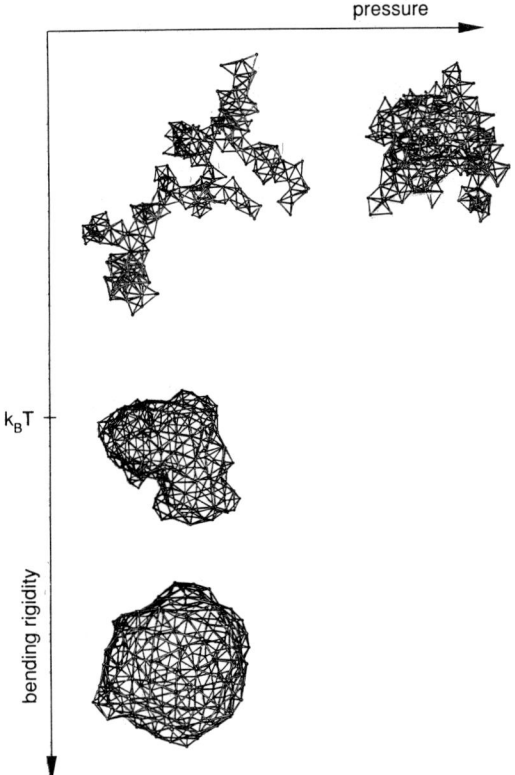

$k_B T$

bending rigidity

Fig. 4.13 Typical configurations of a fluid vesicle with $N = 247$ monomers, for different bending rigidities κ and pressure increments Δp, as indicated. From Kroll and Gompper (1992a), Gompper and Kroll (1992a). Copyright 1992 by the AAAS.

with $\nu_{bp} = 1$ (Parisi and Sourlas, 1981). The same result has also been found for plaquette surfaces on a simple cubic lattice (Cates, 1985; Glaus, 1986, 1988; O'Connell *et al.*, 1991). The spectral dimension d_s, which describes the mean-square displacement of a Brownian particle diffusing in the surface, has also been determined and is found to be $d_s = 1.25 \pm 0.05$, in good agreement with the estimates for branched polymers (Kroll and Gompper, 1992b). The characteristic time, τ_d, it takes a monomer to diffuse over the whole surface is $\tau_d \sim N^{2/d_s}$. However, this is not the longest time scale involved in the simulations. The relaxation time, τ_R, rather scales as $\tau_R \sim N^{1+\nu_{bp}}$ (Gompper and Kroll, 1993).

Fluctuations are suppressed for finite bending rigidity κ. With increasing κ, one finds a smooth crossover from the crumpled to an extended state (Kroll

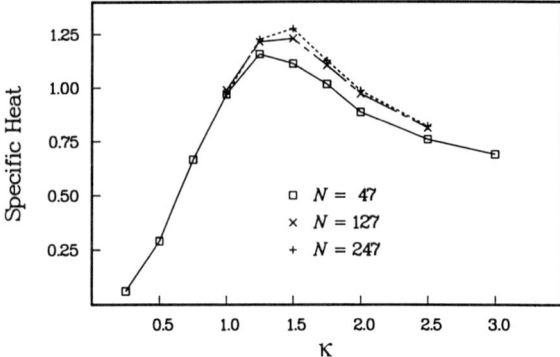

Fig. 4.14 The specific heat of fluid vesicles as a function of the bending rigidity κ. From Kroll and Gompper (1992a). Copyright 1992 by the AAAS.

and Gompper, 1992a; Boal and Rao, 1992a)[†] with a peak in the specific heat when the persistence length reaches the system size, as shown in Fig. 4.14. There do not appear to be distinct large- and small-bending rigidity phases. Thus sufficiently large fluid vesicles will always be crumpled. This is in contrast to many simulations for non-self-avoiding surfaces, in which crumpling transitions have been observed (Catterall, 1989; Baillie et al., 1990, 1991; Renken and Kogut, 1991). However, a recent simulation indicates a lack of phase transition even in the case of a non-self-avoiding membrane (Bowick et al., 1993).

The conformation and scaling behaviour as a function of the pressure increment, for $\kappa = 0$, has also been determined (Gompper and Kroll, 1992a,b). There is a first-order transition between a low-pressure, branched polymer phase to a high-pressure inflated phase. The scaling behaviour in the inflated phase requires the introduction of a new, independent exponent. Arguments similar to those developed by Pincus (1976) and de Gennes (1979), and applied to the study of ring polymers by Camacho and Fisher (1990) and Maggs et al. (1990), imply that the volume V in the inflated phase should scale asymptotically as

$$\langle V \rangle = \Delta p^{3\omega} N^{3\nu_+} \tag{4.82}$$

where

$$\omega = \frac{\nu}{3\nu - 1}$$
$$\nu_+ = \frac{1 - \nu}{3\nu - 1} . \tag{4.83}$$

[†]A different model for fluid membranes has been studied by Drouffe et al. (1991); in this case, only the extended regime is accessible.

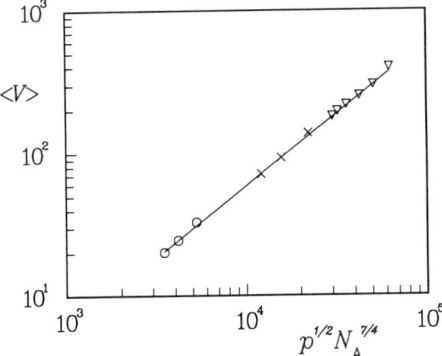

Fig. 4.15 Scaling behaviour of the volume of a fluid vesicle in the inflated phase, as a function of the number of monomers N and the pressure increment Δp. From Gompper and Kroll (1992a).

A scaling analysis of the simulation data is consistent with this behaviour, as seen in Fig. 4.15, and yields the exponent $\nu = 0.79$.

The scaling form (4.82), (4.83) can be understood using a simple generalization of the "blob" picture (Lipowsky and Baumgärtner, 1989; Gompper and Kroll, 1992a,b). Consider a finite piece of a crumpled membrane consisting of N monomers, with the scaling behaviour $\langle R_g^2 \rangle \sim N^\nu$, and assume that a uniform lateral tension Σ is applied to the membrane perimeter. In the blob picture, one envisages the membrane breaking up into N_b blobs of area proportional to the tensile length, ξ_Σ, squared: $\xi_\Sigma^2 = k_B T/\Sigma$. It is then argued that the blobs become independent on length scales much larger than ξ_Σ, so that there are $M_b = N/N_b$ monomers in a blob. One then expects $\xi_\Sigma^2 \simeq a_0^2 M_b^\nu$, where a_0 is a length of the order of the interparticle spacing. The total projected surface area $\langle A_b \rangle$, projected onto the plane in which the tension acts, is therefore proportional to (Gompper and Kroll, 1992a,b)

$$\langle A_b \rangle \sim \frac{N}{M_b} \xi_\Sigma^2 \sim N\Sigma^{1/\nu - 1}. \tag{4.84}$$

An inflated vesicle can now be regarded as a spherical bubble of radius R with a surface tension Σ, with $\Delta p = 2\Sigma/R$. Taking $\langle A_b \rangle \simeq 4\pi R^2$ and $\langle V \rangle \sim R^3$, one arrives at the scaling form (4.82) with the exponents (4.83).

4.8 Level surfaces

The knowledge of the bending modulus of a system is clearly the most important ingredient in the membrane description of self-assembling systems. The idea behind the use of level surfaces is to provide a phenomeno-logical description of the complex pattern of internal interfaces in the microemulsion, and from this pattern extract the bending constants. To do this, the configuration of interface between oil and water regions is associated with a level set of a random standing wave $s_N(\mathbf{r})$ of the form

$$s_N(\mathbf{r}) = \frac{1}{\sqrt{N}} \sum_{i=1}^{N} A_i \cos(\mathbf{k}_i \cdot \mathbf{r} + \varphi_i) \tag{4.85}$$

with uniformly distributed phases φ_i and wavevector directions, and a certain probability distribution $f(k)$ for the magnitude of the wavevector, centred around some preferred value k_0 (Cahn, 1965; Berk, 1987). The levelled density field is then given by

$$\rho(\mathbf{r}) = \Theta_{\alpha\beta}(s_N(\mathbf{r})) \tag{4.86}$$

where $\Theta_{\alpha\beta}(s) = 1$ for $\alpha \leq s \leq \beta$, and 0 otherwise. An alternative to an *a priori* assumption about the probability density $f(k)$ is a natural probability spread in k-space obtained as a consequence of physical localization of the waves in real space. In this case, the random field $s(\mathbf{r})$ has the form (Pieruschka and Marcelja, 1992)

$$s_N(\mathbf{r}) = \frac{1}{\sqrt{N}} \sum_{i=1}^{N} u(|\mathbf{r} - \mathbf{r}_i|; \xi) \cos(\mathbf{k}_i \cdot (\mathbf{r} - \mathbf{r}_i)) \tag{4.87}$$

where $\mathbf{k}_i = k_0 \mathbf{e}_i$ and $\mathbf{e}_i$ is a unit vector. For simplicity, a Yukawa-like localization is chosen, with

$$u(\mathbf{r}; \xi) = \frac{1}{r} (e^{-r/\xi} - e^{-r/a}) \tag{4.88}$$

where a is a constant of the size of the amphiphile length. In the large-N limit, the random standing wave $s_N(\mathbf{r})$ is a Gaussian random field (Adler, 1981; Teubner, 1991). Thus, both the scattering intensity and the averages of the mean curvature H, and the Gaussian curvature K, can be evaluated exactly for the density distribution (4.85), (4.86) (Berk, 1987, 1991; Teubner, 1991). In the case $\beta \to \infty$, one finds (Teubner, 1991)

$$\langle K \rangle = \frac{1}{6} \langle k^2 \rangle (\alpha^2 - 1)$$

$$\langle H \rangle = \frac{\alpha}{2} \sqrt{\frac{\pi}{6} \langle k^2 \rangle}$$

$$\langle H^2 \rangle = \frac{1}{6} \langle k^2 \rangle \left[\alpha^2 + \left(\frac{6}{5} \frac{\langle k^4 \rangle}{\langle k^2 \rangle^2} - 1 \right)^2 \right] . \tag{4.89}$$

Here, $\langle k^2 \rangle$ and $\langle k^4 \rangle$ are moments of the distribution $f(k)$. The average $\langle k^2 \rangle$ is closely related with the surface-to-volume ratio S/V (Berk, 1987, 1991; Teubner, 1991)

$$S/V = \frac{2}{\pi}\exp(-\alpha^2/2)\sqrt{\frac{1}{3}\langle k^2 \rangle} . \tag{4.90}$$

From the above, an expression for the elastic energy, $H_{el}[f]$, is immediately obtained which depends upon the spectral density, $f(k)$, and an unknown bending modulus κ.

To obtain a thermodynamic model for bicontinuous phases, the spectral density $f(k)$ has to be calculated from a free-energy functional (Pieruschka and Marcelja, 1992). To do so, the free energy is written as $F[f] = H_{el}[f] - TS[f]$, where S is the entropy

$$S = \frac{V}{(2\pi)^3}\int_0^\Lambda 4\pi k^2 \ln f(k)\mathrm{d}k. \tag{4.91}$$

The parameter Λ is simply a short-distance cutoff. Independently, structure functions are calculated from (4.87) and (4.88) and fitted to experimentally measured water–water and amphiphile–amphiphile structure functions. These fits determine the parameters k_0 and ξ of the spectral function, $f(k)$. The thermodynamic free energy F is obtained by minimizing $F[f]$ with respect to the coherence length ξ with k_0 fixed at the value given by the fit. This produces a coherence length $\xi(k_0, \kappa)$ which depends on k_0 and on the unknown κ. Finally κ is varied until this value of the coherence length agrees with the value determined by the fit to experimental data. For the SDS system, the scattering intensities have been determined by Auvray et al. (1984). The value of κ obtained by Pieruschka and Marcelja (1992) by fitting these data is about $2k_BT$. Measurements of κ in this system by Safinya et al. (1989) place it between k_BT and $3k_BT$.

In principle, this method provides the means to determine the inputs of the membrane theories, the elastic constants, from some experimental data. The challenge then would be to predict other behaviours of the system from the membrane approach using these inputs, and to compare the results to experiment on the same system.

4.9 Summary

In light of the questions of phase diagrams, interfacial behaviour and structure, how well does the membrane approach do? The few phase diagrams provided at constant temperature, such as that of Fig. 4.8, contain the proper phases but do not look like those of experimental systems, having a

large region of lamellar phase, characteristic of a strong amphiphile, but a
continuous path from oil to water phases, characteristic of a weak one. Phase
diagrams from the more unified approach of Golubović and Lubensky, Fig.
4.9, also show the correct progression of phases, but are not in variables to be
compared with experimental phase diagrams. In this, it is similar to the results
of the Landau–Ginzburg approach, and of most microscopic models. Little
has been obtained from the membrane approach concerning interfacial
properties, because the interfaces are given at the outset. The success of
the approach is certainly in the description of structures. These include
spherical and cylindrical micelles, and the shape and fluctuations of
vesicles, the scaling behaviour of the L_3 phases, and the spacing of lamellar
phases resulting from the Helfrich interaction. This repulsive interaction
clearly tends to enlarge the spacing between membranes, that is, to make
the amphiphiles good solubilizers in a ternary system. As discussed earlier in
Section 2.2.3, there is a direct connection between good solubilizers and low
interfacial tensions. Thus the membrane approach establishes a connection
between thermal fluctuations of the sheets of amphiphile and the low oil/
water interfacial tensions. Finally it provides an understanding of the length
scale which characterizes many amphiphilic structures, the persistence length.
The approach shares with the Landau–Ginzburg theories the shift in emphasis from many microscopic interactions to a few larger-scale parameters
determined by them. In the membrane theories, these parameters are the
elastic constants. The picture of the fluid as an assembly of elastic sheets of
amphiphile is quite different from that of an assembly of interacting
molecules, and lends itself more readily to intuitive analysis of complex
structures. It is this aspect which makes the membrane approach so attractive.

5 Summary and outlook

We have considered three different approaches to the study of amphiphilic
systems. Each of them has advantages and disadvantages.

(i) The microscopic approach, most often formulated on a lattice, has
     produced the only phase diagrams which are readily compared with
     experiment. This is because the phase diagram is sensitive to the degrees
     of freedom of all components which can easily be described in this
     approach. Changes in phase behaviour with the number of hydrophilic
     and hydrophobic groups have been simulated, for example. Reductions
     in interfacial tensions have been calculated, and they are of the correct
     order of magnitude. Such models are also appropriate for the study of
     self-assembly, in that the amphiphilic molecules are free to form an

interface or a micelle as they choose. With the advent of ever more powerful computers, simulation of aggregates and solutions of more realistic molecules will become feasible.

(ii) Ginzburg–Landau models simplify the systems considerably, describing them by means of only a few order parameters and a few coefficients which can be obtained from experiment. Interfacial properties emerge readily from this approach. Because they are continuum models, they describe the structure of ordered phases with curved surfaces very well. The description they provide via simulation of the microemulsion appears to be particularly realistic. They also provide a convenient bridge between the microscopic and the membrane theories.

(iii) Membrane models describe systems in which the amphiphiles are so insoluble that the membranes they form can be treated as entities in themselves. Microscopic differences between the amphiphiles are reflected in the elastic moduli of the membranes. This reduction to a few elastic constants makes it easier to consider changes in shape and conformation. Because the length scale is no longer microscopic, effects of fluctuations arising from very long wavelengths can be studied, and their consequences predicted. Simulations of more complicated arrays of membranes will certainly be carried out in the future.

What is lacking for the most part in all three approaches thus far is quantitative comparison with real systems. All the approaches are capable of fixing their parameters from one kind of experiment and using them to predict the results of completely different experiments. This has not been done as yet, but should be attempted.

Another element lacking is any unified approach to binary and ternary systems. That is, we have argued that these systems are very similar. In the membrane picture, their description is identical. But there have been few attempts to describe a system as it evolves from a balanced ternary one to a binary one as oil, or water, is withdrawn. Clearly the microemulsion evolves into the sponge phase as single layers of amphiphile separating oil and water become binary layers separating inside and outside. This intermediate region in which there are two kinds of internal interface has received little theoretical or experimental attention.

There are several related areas which overlap and stimulate the study of the subjects we have discussed. One of these is the study of ternary polymer mixtures. In them, two different homopolymers consisting of chains of A or B monomers and which normally phase separate can be made compatible on the addition of an AB diblock copolymer (Bates and Fredrickson, 1990; Maglio and Palumbo, 1982). These systems display the same lyotropic phases as exhibited by mixtures of oil, water and amphiphile, as well as forming micelles. These systems are even richer in that the AB copolymer which serves as an amphiphile can be replaced by ABA triblocks which form the analogue of bilayers, or ABC triblocks, or other multiblock assemblies. Mixtures of polymers and amphiphiles present new challenges, as do other mixed systems, such as water insoluble lipids and ordinary amphiphiles, or the system of amphiphiles and impurities.

And then there is the cell membrane with all of its impurities of proteins, and its cytoskeleton which gives the membrane some rigidity. The study of such membranes has been very active, but this takes us too far afield. The above, however, provides a glimpse of related areas of self-assembly which will almost certainly come to fruition in the near future.

6 Acknowledgements

We have been very fortunate to have interacted with many stimulating collaborators, and we would like to acknowledge our debt to them: Gilson Carneiro, Jürgen Goos, Robert Hołyst, Stefan Klein, Martin Kraus, Daniel Kroll, James Lerczak, Mark Matsen, Robert Putz, Friederike Schmid, Wei-Heng Shih and Stefan Zschocke. We have also benefited from numerous conversations with colleagues and friends, to whom we extend our gratitude. Our special thanks go to H. Ted Davis, Eric Kaler, Manfred Kahlweit and Reinhard Strey who have been particularly generous in sharing with us their expertise. Michael Schick is grateful to the Alexander von Humboldt-Stiftung for a Senior Research Scientist award which made possible the visit to the Ludwig-Maximilians Universität München where this review was written, and to Herbert Wagner for his kind hospitality. He would also like to thank the National Science Foundation which has supported his work under past grants, and the present one DMR 9220733. Gerhard Gompper would like to thank Herbert Wagner for stimulating discussions and encouragement. He also acknowledges support from the Deutsche Forschungsgemeinschaft through Sonderforschungsbereich 266.

References

Abillon, O. and Perez, E. (1990). *J. Phys. France* **51**, 2543.
Abillon, O., Lee, L. T., Langevin, D. and Wong, K. (1991). *Physica* **A172**, 209.
Adler, R. J. (1981). *The Geometry of Random Fields*. Wiley, Chichester.
Aharony, A., Domany, E. and Hornreich, R. M. (1987). *Phys. Rev.* **B36**, 2006.
Alexander, S. (1978). *J. Physique Lett.* **39**, L1.
Almdal, K., Rosedale, J. H., Bates, F. S., Wignall, G. D. and Frederickson, G. H. (1990). *Phys. Rev. Lett.* **65**, 1112.

Almdal, K., Koppi, K. A., Bates, F. S. and Mortensen, K. (1992) *Macromolecules* **25**, 1743.
Ambjørn, J., Duurhuus, B. and Fröhlich, J. (1985). *Nucl. Phys.* **B257**, 433.
Ambjørn, J., Duurhuus, B. and Fröhlich, J. (1986). *Nucl. Phys.* **B275**, 161.
Andelman, D., Cates, M. E., Roux, D. and Safran, S. A. (1987). *J. Chem. Phys.* **87**, 7229.
Andersen, G. R. and Wheeler, J. C. (1978). *J. Chem. Phys.* **69**, 2082.
Anderson, D. M., Davis, H. T., Scriven, L. E. and Nitsche, J. C. C. (1990). *Adv. Chem. Phys.* **77**, 337.
Anisimov, M. A., Gorodetsky, E. E., Davydov, A. J. and Kurliansky, S. (1992a). *Mol. Cryst. Liq. Cryst.* **221**, 71.
Anisimov, M. A., Gorodetsky, E. E., Davydov, A. J. and Kurliansky, S. (1992b). *Liquid Crystals* **11**, 941.
Aratono, M. and Kahlweit, M. (1991). *J. Chem. Phys.* **95**, 8578.
Aratono, M. and Kahlweit, M. (1992). *J. Chem. Phys.* **97**, 5932 (E).
Aronowitz, J. A. and Lubensky, T. C. (1988). *Phys. Rev. Lett.* **60**, 2634.
Auvray, L. (1994). *Micelles, Membranes, Microemulsions, and Monolayers* (ed. W. M. Gelbart, D. Roux and A. Ben-Shaul). Springer, Berlin (in press).
Auvray, L., Cotton, J.-P., Ober, R. and Taupin, C. (1984). *J. Phys. Chem.* **88**, 4586.
Auvray, L., Cotton, J.-P., Ober, R. and Taupin, C. (1986). *Physica* **136B**, 281.
Baillie, C. F., Johnston D. A. and Williams, R. D. (1990). *Nucl. Phys.* **B335**, 469.
Baillie, C. F., Caterall, S. M., Johnston D. A. and Williams, R. D. (1991). *Nucl. Phys.* **B348**, 543.
Bak, P. (1982). *Rep. Prog. Phys.* **45**, 587.
Balbuena, P., Borzi, C. and Widom, B. (1986). *Physica* **A138**, 55.
Barber, M. N. (1984). *Phase Transitions and Critical Phenomena*, Vol. 8, p. 145 (ed. C. Domb and J. L. Lebowitz). Academic Press, London.
Bartelt, N. C. and Einstein, T. L. (1986). *J. Phys.* **A19**, 1429.
Bassereau, P. Marignan, J. and Porte, G. (1987). *J. Phys. France* **48**, 673.
Bassereau, P., Marignan, J., Porte, G. and May, R. (1991). *Europhys. Lett.* **15**, 753.
Bates, F. S. and Fredrickson, G. H. (1990). *Ann. Rev. Phys. Chem.* **41**, 525.
Baumgärtner, A. and Ho, J.-S. (1990). *Phys. Rev.* **A41**, 5747.
Beck, J. S., Vartuli, J. C., Roth, W. J., Leonowicz, M. E., Kresge, C. T., Schmitt, K. D., Chu, C. T.-W., Olson, D. H., Sheppard, E. W., McCullen, S. B., Higgins, J. B. and Schlenker, J. L. (1992). *J. Am. Chem. Soc.* **114**, 10834.
Bensimon, D., David, F., Leibler, S. and Pumir, A. (1990). *J. Phys. France* **51**, 689.
Berera, A. and Dawson, K. A. (1990). *Phys. Rev.* **A42**, 3618.
Berera, A. and Kahng, B. (1992). *Phys. Rev.* **A46**, 4528.
Berk, N. F. (1987). *Phys. Rev. Lett.* **58**, 2718.
Berk, N. F. (1991). *Phys. Rev.* **A44**, 5069.
Berndl, K., Käs, J., Lipowsky, R., Sackmann, E. and Seifert, U. (1990). *Europhys. Lett.* **13**, 659.
Billoire, A. and David, F. (1986). *Nucl. Phys.* **B275**, 617.
Binks, B. P., Kellay, H. and Meunier, J. (1991). *Europhys. Lett.* **16**, 53.
Bivas, I., Hanusse, P., Bothorel, P., Lalanne, J. and Aguerre-Chariol, O. (1987). *J. Phys. (Paris)* **48**, 855.
Blokhuis, E. M. and Bedeaux, D. (1991). *J. Chem. Phys.* **95**, 6986.
Blokhuis, E. M. and Bedeaux, D. (1992). *Physica* **A184**, 42.
Blokhuis, E. M. and Bedeaux, D. (1993). *Mol. Phys.* **80**, 705.
Bloom, M. and Evans, E. (1991). *Biologically Inspired Physics* (ed. L. Peliti). Plenum Press, New York.

Blossey, R. and Schick, M. (1991). *Phys. Rev.* A**44**, 1134.
Blume, M., Emery, V. and Griffiths, R. B. (1971). *Phys. Rev.* A**4**, 1071.
Boal, D. and Rao, M. (1992a). *Phys. Rev.* A**45**, R6947.
Boal, D. and Rao, M. (1992b). *Phys. Rev.* A**46**, 3037.
Bodet, J.-F., Bellare, J. R., Davis, H. T., Scriven, L. E. and Miller, W. G. (1988a). *J. Phys. Chem.* **92**, 1898.
Bodet, J-F., Davis, H. T., Scriven, L. E. and Miller, W. G. (1988b). *Langmuir* **4**, 455.
Borzi, C., Lipowsky, R. and Widom, B. (1986), *J. Chem. Soc., Faraday Trans. 2*, **82**, 1752.
Boulatov, D. V., Kazakov, V. A, Kostov, I. K. and Migdal, A. A. (1986). *Nucl. Phys.* B**275**, 641.
Bowick, M., Coddington, P., Han, L., Harris, G. and Marinari, E. (1993). *Nucl. Phys.* B**394**, 791.
Brazovskii, S. A. (1975). *Zh. Eksp. Teor. Fiz.* **68**, 175. (*Sov. Phys. JETP* **41**, 85.)
Brochard, F. and Lennon, J. F. (1975). *J. Phys. (Paris)* **36**, 1035.
Bruinsma, R. (1992). *J. Phys. II France* **2**, 425.
Cahn, J. W. (1965). *J. Chem. Phys.* **42**, 93.
Cahn, J. W. (1977). *J. Chem. Phys.* **66**, 3667.
Caillé, A. (1972). *C.R. Acad. Sci. Ser. B* **274**, 891.
Camacho, C. J. and Fisher, M. E. (1990). *Phys. Rev. Lett.* **65**, 9.
Canham, P. B. (1970). *J. Theor. Biol.* **26**, 61.
Cappi, A., Colangelo, P., Gonella, G. and Maritan, A. (1992). *Nucl. Phys.* B**370**, 659.
Carneiro, G. M. and Schick, M. (1988). *J. Chem. Phys.* **89**, 4638.
Cates, M. E. (1985). *Phys. Lett.* B**161**, 363.
Cates, M. E., Andelman, D., Safran, S. A. and Roux, D. (1988a). *Langmuir* **4**, 802.
Cates, M. E., Roux, D., Andelman, D., Milner, S. T. and Safran, S. A. (1988b). *Europhys. Lett.* **5**, 733.
Catterall, S. M. (1989). *Phys. Lett.* B**220**, 207.
Cevc, G., Fenzl, W. and Sigl, L. (1990). *Science* **249**, 1161.
Chandra, P. and Safran, S. A. (1992), *Europhys. Lett.* **17**, 691.
Charvolin, J. (1990). *Contemp. Phys.* **31**, 1.
Charvolin, J. and Sadoc, J. F. (1987). *J. Phys. (Paris)* **48**, 1559.
Chen, K., Ebner, C., Jayaprakash, C. and Pandit, R. (1987). *J. Phys.* C**20**, L361.
Chen, K., Ebner, C., Jayaprakash, C. and Pandit, R. (1988). *Phys. Rev.* A**38**, 6240.
Chen, K., Jayaprakash, C., Pandit, R. and Wenzel, W. (1990). *Phys. Rev. Lett.* **65**, 2736.
Chen, L.-J., Jeng, J.-F., Robert M. and Shukla, K. P. (1990). *Phys. Rev.* A**42**, 1821.
Chen, S.-H., Chang, S.-L. and Strey, R. (1990). *J. Chem. Phys.* **93**, 1907.
Chen, S.-H., Chang, S.-L., Strey, R., Samseth, J. and Mortensen, K. (1991). *J. Phys. Chem.* **95**, 7427.
Chen, S.-H., Huang, J. S. and Tartaglia, P. (eds) (1992). *Structure and Dynamics of Strongly Interacting Colloids and Supramolecular Aggregates in Solution*. Kluwer Academic Publishers, Dordrecht.
Chowdhury, D. and Stauffer, D. (1991). *Phys. Rev.* A**44**, 2247.
Ciach, A. (1990). *J. Chem. Phys.* **93**, 5322.
Ciach, A. (1992). *J. Chem. Phys.* **96**, 1399.
Ciach, A. and Høye, J. S. (1990). *J. Chem. Phys.* **90**, 1222.
Ciach, A., Høye, J. S. and Stell, G. (1988). *J. Phys.* A**21**, L777.
Ciach, A., Høye, J. S. and Stell, G. (1989). *J. Chem. Phys.* **90**, 1214.
Ciach, A., Høye, J. S. and Stell, G. (1991). *J. Chem. Phys.* **95**, 5300.

Colangelo, P., Gonella, G. and Maritan, A. (1993). *Phys. Rev.* E**47**, 411.
Coulon, C., Roux, D. and Bellocq, A. M. (1991). *Phys. Rev. Lett.* **66**, 1709.
David, F. (1986). *Europhys. Lett.* **2**, 577.
David, F. (1989). *Statistical Mechanics of Membranes and Surfaces* (ed. D. Nelson, T. Piran and S. Weinberg). World Scientific, Singapore.
David, F. (1990). *J. Phys. France* **51** (Colloque), C7–115.
David, F. and Guitter, E. (1987). *Europhys. Lett.* **3**, 1169.
David, F. and Leibler, S. (1991). *J. Phys. II France* **1**, 959.
Davis, H. T., Bodet, J. F., Scriven, L. E. and Miller, W. G. (1987). *Physics of Amphiphilic Layers* (ed. J. Meunier, D. Langevin and N. Boccara). Springer, Berlin.
Dawson, K. A. (1987). *Phys. Rev.* A**35**, 1766.
Dawson, K. A. and Kurtović, Z. (1990). *J. Chem. Phys.* **92**, 5473.
Dawson, K. A., Lipkin, M. D. and Widom, B. (1988). *J. Chem. Phys.* **88**, 5149.
Dawson, K. A. Walker, B. L. and Berera, A. (1990). *Physica* A**165**, 320.
Debye, P., Anderson, H. R. Jr. and Brumberger, H. (1957). *J. Appl. Phys.* **28**, 679.
de Gennes, P. G. (1974). *The Physics of Liquid Crystals*. Clarendon, Oxford.
de Gennes, P. G. (1979). *Scaling Concepts in Polymer Physics*. Cornell University Press, Ithaca, New York.
de Gennes, P. G. and Taupin, C. (1982). *J. Phys. Chem.* **86**, 2294.
Degiorgio, V., Corti, M. and Cantú, L. (1988). *Chem. Phys. Lett.* **151**, 349.
Derrick, G. H. (1964). *J. Math. Phys.* **5**, 1252.
Deuling, H. J. and Helfrich, W. (1976). *J. Phys. (Paris)* **37**, 1335.
Diehl, H.-W., Kroll, D. M. and Wagner, H. (1980). *Z. Phys.* B**36**, 329.
Dietrich, S. (1988). *Phase Transitions and Critical Phenomena*, Vol. 12 (ed. C. Domb and J. Lebowitz). Academic Press, London.
di Meglio, J. M. and Bassereau, P. (1991). *J. Phys. II France* **1**, 247.
di Meglio, J. M., Dvolaitzky, M., Leger, L. and Taupin, C. (1985). *Phys. Rev. Lett.* **54**, 1686.
Drouffe, J.-M., Maggs, A. C. and Leibler, S. (1991). *Science* **254**, 1353.
Duplantier, B., Goldstein, R. E., Romero-Rochin, V. and Pesci, A. I. (1990). *Phys. Rev. Lett.* **65**, 508.
Duwe, H.-P., Engelhardt, H., Zilker, A. and Sackmann, E. (1987). *Mol. Cryst. Liq. Cryst.* **152**, 1.
Duwe, H.-P., Zeman, K. and Sackmann, E. (1989). *Prog. Colloid Polym. Sci.* **79**, 6.
Duwe, H.-P., Käs, J. and Sackmann, E. (1990). *J. Phys. France* **51**, 945.
Elgsaeter, A., Stokke, B., Mikkelsen, A. and Branton, D. (1986). *Science* **234**, 1217.
Evans, E. (1974). *Biophys. J.* **14**, 923.
Evans, E. and Rawicz, W. (1990). *Phys. Rev. Lett.* **64**, 2094.
Farago, B., Richter, D., Huang, J. S., Safran, S. A. and Milner, S. T. (1990). *Phys. Rev. Lett.* **65**, 3348.
Faucon, J. F., Mitov, M. D., Meleard, P., Bivas, I. and Bothorel, P. (1989). *J. Phys. (Paris)* **50**, 2389.
Firman, P., Haase, D., Jen, J., Kahlweit, M. and Strey, R. (1985). *Langmuir* **1**, 718.
Fisher, M. E. and Fisher, D. (1982). *Phys. Rev.* B**25**, 3102.
Fisher, M. E. and Huse, D. A. (1982). *Melting, Localization, and Chaos*, p. 259 (ed. R. K. Kalia and P. Vashista). Elsevier, New York.
Fisher, M. E. and Jin, A. J. (1992). *Phys. Rev. Lett.* **69**, 792.
Fisher, M. E. and Widom, B. (1969). *J. Chem. Phys.* **50**, 3756.
Fisher, M. P. A. and Wortis, M. (1984). *Phys. Rev.* B**29**, 6252.

Forgacs, G., Nieuwenhuizen, T. M. and Lipowsky, R. (1991). *Phase Transitions and Critical Phenomena*, Vol. 14 (ed. C. Domb and J. Lebowitz). Academic Press, London.
Förster, D. (1986). *Phys. Lett.* A114, 115.
Furman, D., Dattagupta, S. and Griffiths, R. B. (1977). *Phys. Rev.* B15, 441.
Gelbart, W. M., Roux, D. and Ben-Shaul, A. (eds) (1994). *Micelles, Membranes, Microemulsions, and Monolayers.* Springer, Berlin (in press).
Gibbs, J. W. (1878). *Trans. Conn. Acad.* III, 343; reprinted (1948) *The Collected Works of J. Willard Gibbs*, Vol. I, p.258. Yale, New Haven.
Glaus, U. (1986). *Phys. Rev. Lett.* 56, 1996.
Glaus, U. (1988). *J. Stat. Phys.* 50, 1141.
Goldstein, R. E. (1985). *J. Chem. Phys.* 83, 1246.
Goldstein, R. E. (1986). *J. Chem. Phys.* 84, 3367.
Golubović, L. and Lubensky, T. C. (1989a). *Phys. Rev.* B39, 12110.
Golubović, L. and Lubensky, T. C. (1989b). *Europhys. Lett.* 10, 513.
Golubović, L. and Lubensky, T. C. (1990). *Phys. Rev.* A41, 4343.
Gompper, G. (1992). *Structure and Dynamics of Strongly Interacting Colloids and Supramolecular Aggregates in Solution* (ed. S.-H. Chen, J. S. Huang, and P. Tartaglia). Kluwer Academic Publishers, Dordrecht.
Gompper, G. and Klein, S. (1992). *J. Phys. II France* 2, 1725.
Gompper, G. and Kraus, M. (1993a). *Phys. Rev.* E47, 4289.
Gompper, G. and Kraus, M. (1993b). *Phys. Rev.* E47, 4301.
Gompper, G. and Kroll, D. M. (1989). *Europhys. Lett.* 9, 59.
Gompper, G. and Kroll, D. M. (1991). *Europhys. Lett.* 15, 783.
Gompper, G. and Kroll, D. M. (1992a). *Europhys. Lett.* 19, 581.
Gompper, G. and Kroll, D. M. (1992b). *Phys. Rev.* A46, 7466.
Gompper, G. and Kroll, D. M. (1993). *Phys. Rev. Lett.* 71, 1111.
Gompper, G. and Schick, M. (1989a). *Phys. Rev. Lett.* 62, 1647.
Gompper, G. and Schick, M. (1989b). *Chem. Phys. Lett.* 163, 475.
Gompper, G. and Schick, M. (1990a). *Phys. Rev.* B41, 9148.
Gompper, G. and Schick, M. (1990b). *Phys. Rev.* A42, 2137.
Gompper, G. and Schick, M. (1990c). *Phys. Rev. Lett.* 65, 1116.
Gompper, G. and Schick, M. (1994a). *Micelles, Membranes, Microemulsions, and Monolayers* (ed. W. M. Gelbart, D. Roux and A. Ben-Shaul). Springer, Berlin (in press).
Gompper, G. and Schick, M. (1994b). *Phys. Rev.* E49, 1491.
Gompper, G. and Zschocke, S. (1991). *Europhys. Lett.* 16, 731.
Gompper, G. and Zschocke, S. (1992a). *The Structure and Conformation of Amphiphilic Membranes* (ed. R. Lipowsky, D. Richter and K. Kremer). Springer, Berlin.
Gompper, G. and Zschocke, S. (1992b). *Phys. Rev.* A46, 4836.
Gompper, G. Hołyst, R. and Schick, M. (1991). *Phys. Rev.* A43, 3157.
Goos, J. and Gompper, G. (1993). *J. Phys. I France* 3, 1551.
Gösmann, K. (1990). *Diplomarbeit*, Universität München.
Gruner, S. M. (1989). *J. Phys. Chem.* 93, 7562.
Guéring, P. and Lindman, B. (1985). *Langmuir* 1, 464.
Gunn, J. R. and Dawson, K. A. (1989). *J. Chem. Phys.* 91, 6393.
Gunn, J. R. and Dawson, K. A, (1992). *J. Chem. Phys.* 96, 3152.
Hackenbroich, G. (1988), *Diplomarbeit*, Universität München.
Hadziioannou, G. and Skoulios, A. (1982). *Macromolecules* 15, 258.

Halley, J. W. and Kolan, A. J. (1988). *J. Chem. Phys.* **88**, 3313.
Hansen, A. Schick, M. and Stauffer, D. (1991). *Phys. Rev.* A**44**, 3686.
Harbich, W., Servuss, R. M. and Helfrich, W. (1978). *Z. Naturforsch.* **33a**, 1013.
Helfrich, W. (1973). *Z. Naturforsch.* **28c**, 693.
Helfrich, W. (1974). *Phys. Lett.* A**50**, 115.
Helfrich, W. (1978). *Z. Naturforsch.* **33a**, 305.
Helfrich, W. (1981). *Physics of Defects* (ed. R. Balian, M. Kléman, and J.-P. Poirier). North Holland, Amsterdam.
Helfrich, W. (1985). *J. Phys. France* **46**, 1263.
Hirschfelder, J. D., Stevenson, D. and Eyring, H. (1937). *J. Chem. Phys.* **5**, 896.
Ho, J.-S. and Baumgärtner, A. (1990). *Europhys. Lett.* **12**, 295.
Hoffmann, H. (1990). *Adv. Colloid Interface Sci.* **32**, 123.
Hofsäss, T. and Kleinert, H. (1988). *J. Chem. Phys.* **88**, 1156.
Hornreich, R. M., Luban, M. and Shtrikman, S. (1975). *Phys. Rev. Lett.* **35**, 1678.
Huse, D. A. and Leibler, S. (1988). *J. Phys. France* **49**, 605.
Huse, D. A. and Leibler, S. (1991). *Phys. Rev. Lett.* **66**, 437.
Israelachvili, J. N. and Adams, G. E. (1978). *Chem. Soc. Faraday Trans.* I **74**, 975.
Jahn, W. and Strey, R. (1988). *J. Chem. Phys.* **92**, 2294.
Jan, N. and Stauffer, D. (1988). *J. Phys. France* **49**, 623.
Janke, W. (1990). *Int. J. Mod. Phys.* B**4**, 1763.
Janke, W. and Kleinert, H. (1986). *Phys. Lett.* A**117**, 353.
Janke, W. and Kleinert, H. (1987). *Phys. Rev. Lett.* **58**, 144.
Janke, W., Kleinert, H. and Meinhard, M. (1989). *Phys. Lett.* B**217**, 525.
Kahlweit, M., Strey, R., Firman, P. and Haase, D. (1985). *Langmuir* **1**, 281.
Kahlweit, M., Strey, R. and Firman, P. (1986). *J. Phys. Chem.* **90**, 671.
Kahlweit, M., Strey, R., Haase, D., Kuneida, H., Schmeling, T., Faulhaber, B., Borkovec, M., Eicke, H.-F., Busse, G., Eggers, F., Funck, T., Richmann, H., Magid, L., Söderman, O., Stilbs, P., Winkler, J., Dittrich, A. and Jahn, W. (1987). *J. Colloid Interface Sci.* **118**, 436.
Kahlweit, M., Strey, R. and Busse, G. (1990). *J. Phys. Chem.* **94**, 3881.
Kahlweit, M., Strey, R., Aratono, M., Busse, G., Jen, J. and Schubert, K.-V. (1991). *J. Chem. Phys.* **95**, 2842.
Kahlweit, M., Strey, R. and Busse, G. (1993). *Phys. Rev.* E**47**, 4197.
Kahng, B., Berera, A. and Dawson, K. A. (1990). *Phys. Rev.* A**42**, 6093.
Kaler, E. W., Murthy, A. K., Rodriguez, B. E. and Zasadzinski, J. A. N. (1989). *Science* **245**, 1371.
Kantor, Y. and Nelson, D. R. (1987a). *Phys. Rev. Lett.* **58**, 2774.
Kantor, Y. and Nelson, D. R. (1987b). *Phys. Rev.* A**36**, 4020.
Kantor, Y., Kardar, M. and Nelson, D. R. (1986). *Phys. Rev. Lett.* **57**, 791.
Kantor, Y., Kardar, M. and Nelson, D. R. (1987). *Phys. Rev.* A**35**, 3056.
Kasteleyn, P.W. and Fortuin, C. M. (1969). *J. Phys. Soc. Japan (suppl.)* **26**, 11.
Kawakatsu, T. and Kawasaki, K. (1990). *Physica* A**167**, 690.
Kawasaki, K. and Kawakatsu, T. (1990). *Physica* A**164**, 549.
Kékicheff, P. and Christenson, H. K. (1989). *Phys. Rev. Lett.* **63**, 2823.
Kellay, H., Meunier, J. and Binks, B. P. (1993). *Phys. Rev. Lett.* **70**, 1485.
Keller, J. B. and Merchant, G. J. (1991). *J. Stat. Phys.* **63**, 1039.
Kleinert, H. (1986a). *J. Chem. Phys.* **84**, 964.
Kleinert, H. (1986b). *Phys. Lett.* A**114**, 263.
Knickerbocker, B. M., Pesheck, C. V., Davis, H. T. and Scriven, L. E. (1982). *J. Phys. Chem.* **86**, 393.

Kroll, D. M. and Gompper, G. (1992a). *Science* **255**, 968.

Kroll, D. M. and Gompper, G. (1992b). *Phys. Rev.* A**46**, 3119.

Kroll, D. M. and Gompper, G. (1993). *Statistical Thermodynamics and Differential Geometry of Microstructure Materials* (ed. H. Ted Davis and Johannes C. C. Nitsche). Springer-Verlag, New York (in press).

Kummrow, M. and Helfrich, W. (1991). *Phys. Rev.* A**44**, 8356.

Lagues, M. and Sauterey, C. (1980). *J. Phys. Chem.* **84**, 3503.

Landau, L. D. (1965). *Collected Papers*, p.209. Pergamon Press, Oxford.

Lang, J. C. and Morgan, R. D. (1980). *J. Chem. Phys.* **73**, 5849.

Laradji, M., Guo, H., Grant, M. and Zuckermann, M. J. (1991a). *Phys. Rev.* A**44**, 8184.

Laradji, M., Guo, H., Grant, M. and Zuckermann, M. J. (1991b). *J. Phys.* A**24**, L629.

Laradji, M., Guo, H., Grant, M. and Zuckermann, M. J. (1992). *J. Phys. Condens. Matter* **4**, 6715.

Larche, F. C., Appell, J., Porte, G., Bassereau, P. and Marignan, J. (1986). *Phys. Rev. Lett.* **56**, 1700.

Larson, R. G. (1988). *J. Chem. Phys.* **89**, 1642.

Larson, R. G. (1989). *J. Chem. Phys.* **91**, 2479.

Larson, R. G. (1992). *J. Chem. Phys.* **96**, 7904.

Larson, R. G., Scriven, L. E. and Davis, H. T. (1985). *J. Chem. Phys.* **83**, 2411.

Lee, L. T., Langevin, D. and Farnoux, B. (1991). *Phys. Rev. Lett.* **67**, 2678.

Leibler, S. and Lipowsky, R. (1987). *Phys. Rev.* B**35**, 7004.

Lekkerkerker, H. N. W. (1989). *Physica* A**159**, 319.

Lekkerkerker, H. N. W. (1990). *Physica* A**167**, 384.

Lerczak, J., Schick, M. and Gompper, G. (1992). *Phys. Rev.* A**46**, 985.

Levin, Y. and Dawson, K. A. (1990). *Phys. Rev.* A**42**, 1976.

Levin, Y., Mundy, C. J. and Dawson, K. A. (1992). *Phys. Rev.* A**45**, 7309.

Lifshitz, E. M. and Pitaevskii, L. P. (1980). *Statistical Physics*, 3rd Ed. Part 1, Sec. 145. Pergamon Press, Oxford.

Lin, S. C. and Lowe, M. J. (1983). *J. Phys.* A**16**, 347.

Lipowsky, R. (1984). *Z. Phys.* B**55**, 345.

Lipowsky, R. (1991). *Nature* **349**, 475.

Lipowsky, R. and Baumgärtner, A. (1989). *Phys. Rev.* A**40**, 2078.

Lipowsky, R. and Fisher, M. E. (1986). *Phys. Rev. Lett.* **57**, 2411.

Lipowsky, R. and Leibler, S. (1986). *Phys. Rev. Lett.* **56**, 2541.

Lipowsky, R. and Leibler, S. (1987). *Phys. Rev. Lett.* **59**, 1983 (E).

Lipowsky, R., Richter, D. and Kremer, K. (eds) (1992). *The Structure and Conformation of Amphiphilic Membranes*. Springer, Berlin.

Litster, J. D. (1975). *Phys. Lett.* A**53**, 193.

Maggs, A. C., Leibler, S., Fisher, M. E. and Camacho, C. J. (1990). *Phys. Rev.* A**42**, 691.

Maglio, G. and Palumbo, R. (1982). *Polymer Blends* (eds M. Kryszewski, A. Galeski and E. Martuscelli). Plenum, New York.

Matsen, M. W. and Schick, M. (1993). *Macromolecules* **26**, 3878.

Matsen, M. W. and Sullivan, D. E. (1990). *Phys. Rev.* A**41**, 2021.

Matsen, M. W. and Sullivan, D. E. (1991). *Phys. Rev.* A**44**, 3710.

Matsen, M. W. and Sullivan, D. E. (1992a). *J. Phys. II France* **2**, 93.

Matsen, M. W. and Sullivan, D. E. (1992b). *Phys. Rev.* A**46**, 1985.

Matsen, M. W. Schick, M. and Sullivan, D. E. (1993). *J. Chem. Phys.* **98**, 2341.

Mecke, K. (1989). *Diplomarbeit*, Universität München.

Méléard, P., Faucon, J. F., Mitov, M. D. and Bothorel, P. (1992). *Europhys. Lett.* **19**, 267.

Meunier, J. (1985). *J. Phys. (Paris) Lett.* **46**, L1005.
Meunier, J. and Lee, L. T. (1991). *Langmuir* **7**, 1855.
Meunier, J., Langevin, D. and Boccara, N. (eds) (1987). *Physics of Amphiphilic Layers.* Springer, Berlin.
Miao, L., Fourcade, B., Rao, M., Wortis, M. and Zia, R. K. P. (1991). *Phys. Rev.* **A43**, 6843.
Milner, S. T. and Safran, S. A. (1987). *Phys. Rev.* **A36**, 4371.
Milner, S. T. and Witten, T. A. (1988). *J. Phys. France* **49**, 1951.
Milner, S. T., Safran, S. A., Andelman, D., Cates, M. E. and Roux, D. (1988). *J. de Physique* **49**, 1065.
Minchau, B., Dünweg, B. and Binder, K. (1990). *Polym. Comm.* **31**, 348.
Mitchell, D. J., Tiddy, G. J., Waring, L., Bostock, T. and McDonald, M. P. (1983). *J. Chem. Soc. Faraday Trans. 1*, **79**, 975.
Morawietz, D., Chowdhury, D., Vollmar, S. and Stauffer, D. (1992). *Physica* **A187**, 126.
Müller-Hartmann, E. and Zittartz, J. (1977). *Z. Phys.* **B27**, 261.
Mukamel, D. and Blume, M. (1974). *Phys. Rev.* **A10**, 610.
Murata, K. K. (1979). *J. Phys.* **A12**, 81.
Mutz, M. and Helfrich, W. (1989). *Phys. Rev. Lett.* **62**, 2881.
Nallet, F., Roux, D. and Milner, S. T. (1990). *J. Phys. France* **51**, 2333.
Napiorkowski, M. and Dietrich, S. (1992). *Z. Phys.* **B89**, 263.
Napiorkowski, M. and Dietrich, S. (1993). *Phys. Rev.* **E47**, 1836.
Nelson, D. R. (1989). *Statistical Mechanics of Membranes and Surfaces* (eds D. Nelson, T. Piran and S. Weinberg). World Scientific, Singapore.
Nelson, D. R. and Peliti, L. (1987), *J. Phys. France* **48**, 1085.
Nelson, D., Piran, T. and Weinberg, S. (eds) (1989). *Statistical Mechanics of Membranes and Surfaces.* World Scientific, Singapore.
Nightingale, M. P. (1976). *Physica* **A83**, 561.
O'Connell, J., Sullivan, F., Libes, D., Orlandini, E., Tesi, M. C., Stella, A. L. and Einstein, T. L. (1991). *J. Phys.* **A24**, 4619.
Olsson, U., Shinoda, K. and Lindman, B. (1986). *J. Phys. Chem.* **90**, 4083.
Paczuski, M., Kardar, M. and Nelson, D. R. (1988). *Phys. Rev. Lett.* **60**, 2638.
Pandit, R., Schick, M. and Wortis, M. (1982). *Phys. Rev.* **B26**, 5112.
Parisi, G. and Sourlas, N. (1981). *Phys. Rev. Lett.* **46**, 871.
Peierls, R. E. (1934). *Helv. Phys. Acta* **7** Suppl., 81.
Peliti, L. and Leibler, S. (1985). *Phys. Rev. Lett.* **54**, 1690.
Pershan, P. S. (1989). *Coll. Phys.* **50**, C7-1.
Peterson, M. A. (1985a). *J. Appl. Phys.* **57**, 1739.
Peterson, M. A. (1985b). *J. Math. Phys.* **26**, 711.
Peyrelasse, J. and Boned, C. (1990). *Phys. Rev.* **A41**, 938.
Pieruschka, P. and Marcelja, S. (1992). *J. Phys. II France* **2**, 235.
Pincus, P. (1976). *Macromolecules* **9**, 386.
Pincus, P., Joanny, J.-F. and Andelman, D. (1990). *Europhys. Lett.* **11**, 763.
Plischke M. and Boal, D. (1988). *Phys. Rev.* **A38**, 4943.
Polyakov, A. M. (1986). *Nucl. Phys.* **B268**, 406.
Polyakov, A. M. (1988). *Gauge Fields and Strings.* Harwood, New York.
Porod, G. (1951). *Kolloid Z.* **124**, 83.
Porte, G. (1992). *J. Phys. Cond. Matter* **4**, 8649.
Porte, G., Marignan, J., Bassereau, P. and May, R. (1988a). *Europhys. Lett.* **7**, 713.
Porte, G., Marignan, J., Bassereau, P. and May, R. (1988b). *J. Phys. France* **49**, 511.

Porte, G., Appell, J., Bassereau, P. and Marignan, J. (1989). *J. Phys. France* **50**, 1335.

Porte, G., Delsanti, M., Billard, I., Skouri, M., Appell, J., Marignan, J. and Debeauvais, F. (1991). *J. Phys. II France* **1**, 1101.

Pouchelon, A., Meunier, J., Langevin, D., Chatenay, D. and Cazabat, A. M. (1980). *Chem. Phys. Lett.* **76**, 277.

Putz, J., Hołyst, R. and Schick, M. (1992). *Phys. Rev.* A**46**, 3369.

Putz, J., Hołyst, R. and Schick, M. (1993). *Phys. Rev.* E**48**, 635 (E).

Renken, R. L. and Kogut, J. B. (1991). *Nucl. Phys.* B**354**, 328.

Renlie, L., Høye, J. S., Skaf, M. S. and Stell, G. (1991). *J. Chem. Phys.* **95**, 5305.

Richetti, P., Kékicheff, P., Parker, J. L. and Ninham, B. W. (1990). *Nature* **346**, 252.

Robert, M. and Jeng, J. F. (1988). *J. Phys. France* **49**, 1821.

Romero-Rochin, V., Varea, C. and Robledo, A. (1991). *Phys. Rev.* A**44**, 8417.

Romero-Rochin, V., Varea, C. and Robledo, A. (1992). *Physica* A**184**, 367.

Rosevear, F. B. (1968). *J. Soc. Cosmetic Chemists* **19**, 581.

Roux, D. and Safinya, C. R. (1988). *J. Phys. France* **49**, 307.

Roux, D., Cates, M. E., Olsson, U., Ball, R. G., Nallet, F. and Bellocq, A. M. (1990). *Europhys. Lett.* **11**, 229.

Roux, D., Coulon, C. and Cates, M. E. (1992a). *J. Phys. Chem.* **96**, 4174.

Roux, D., Nallet, F., Freyssingeas, E., Porte, G., Bassereau, P., Skouri, M. and Marignan, J. (1992b). *Europhys. Lett.* **17**, 575.

Rowlinson, J. S. and Widom, B. (1982). *Molecular Theory of Capillarity*. Clarendon Press, Oxford.

Sackmann, E. (1982). *Biophysics* (ed. W. Hoppe, W. Lohmann, H. Markl and H. Ziegler). Springer, Berlin.

Safinya, C., Roux, D., Smith, G., Sinha, S., Dimon, P., Clark, A. and Bellocq, A. (1986). *Phys. Rev. Lett.* **57**, 2718.

Safinya, C. R., Sirota, E. B., Roux, D. and Smith, G. S. (1989). *Phys. Rev. Lett.* **62**, 1134.

Safran, S. A. (1983). *J. Chem. Phys.* **78**, 2073.

Safran, S. A. (1991). *Phys. Rev.* A**43**, 2903.

Safran, S. A., Turkevich, L. A. and Pincus, P. A. (1984). *J. Phys. (Paris) Lett.* **45**, L69.

Safran, S. A., Roux, D., Cates, M. E. and Andelman, D. (1986). *Phys. Rev. Lett.* **57**, 491.

Safran, S. A., Pincus, P. A. and Andelman, D. (1990). *Science* **248**, 354.

Safran, S. A., Pincus, P. A., Andelman, D. and MacKintosh, F. C. (1991). *Phys. Rev.* A**43**, 1071.

Schick, M. (1990). *Liquids at Interfaces* (ed. J. Charvolin, J. F. Joanny and J. Zinn-Justin). North-Holland, Amsterdam.

Schick, M. and Shih, W.-H. (1986). *Phys. Rev.* B**34**, 1797.

Schick, M. and Shih, W.-H. (1987). *Phys. Rev. Lett.* **59**, 1205.

Schmid, F. and Schick, M. (1994). *Phys. Rev.* E**49**, 494.

Schneider, M. D., Jenkins, J. T. and Webb, W. W. (1984). *J. Phys. (Paris)* **45**, 1457.

Schubert, K.-V. and Strey, R. (1991). *J. Chem. Phys.* **95**, 8532.

Scriven, L. E. (1976). *Nature* **263**, 123.

Scriven, L. E. (1977). *Micellization, Solubilization, and Microemulsions* (ed. K. L. Mittal). Plenum Press, New York.

Seddon, J. M. (1990). *Biochim. Biophys. Acta* **1031**, 1.

Seeto, Y., Puig, J. E., Scriven, L. E. and Davis, H. T. (1983). *J. Colloid Interface Sci.* **96**, 360.

Seifert, U., Berndl, K. and Lipowsky, R. (1991). *Phys. Rev.* A**44**, 1182.

Selke, W. (1988). *Physics Rep.* **170**, 213.
Selke, W. (1992). *Phase Transitions and Critical Phenomena*, Vol. 15 (ed. C. Domb and J. Lebowitz). Academic, New York.
Sivardiere, J. and Lazerowicz, J. (1975). *Phys. Rev.* **A11**, 2090.
Skaf, M. S. and Stell, G. (1992a). *J. Chem. Phys.* **97**, 7699.
Skaf, M. S. and Stell, G. (1992b). *Phys. Rev.* **A46**, R3020.
Skaf, M. S. and Stell, G. (1993). *J. Phys.* **A51**, 1051.
Skouri, M., Marignan, J., Appell, J. and Porte, G. (1991). *J. Phys. II France* **1**, 1121.
Slotte, P. A. (1992). *Phys. Rev.* **A46**, 6469.
Smit, B., Hilbers, P. A. J., Esselink, K., Rupert, L. A. M., van Os, N. M. and Schlijper, A. G. (1990). *Nature* **348**, 624.
Smit, B., Hilbers, P. A. J., Esselink, K., Rupert, L. A. M., van Os, N. M. and Schlijper, A. G. (1991). *J. Phys. Chem.* **95**, 6361.
Smit, B., Esselink, K., Hilbers, P. A. J., van Os, N. M., Rupert, L. A. M. and Szleifer, I. (1993). *Langmuir* **9**, 9.
Smith, D. H. and Covatch, G. L. (1990). *J. Chem. Phys.* **93**, 6870.
Stauffer, D. (1985). *Introduction to Percolation Theory*. Taylor and Francis, London.
Stauffer, D. and Eicke, H. F. (1992). *Physica* **A182**, 29.
Stauffer, D. and Jan, N. (1987). *J. Chem. Phys.* **87**, 6210.
Stephenson, J. (1970). *J. Math. Phys.* **11**, 420.
Stockfisch, T. P. and Wheeler, J. C. (1988). *J. Phys. Chem.* **92**, 3292.
Stockfisch, T. P. and Wheeler, J. C. (1993). *J. Chem. Phys.* **99**, 6155.
Stohrer, J., Gröbner, G., Reimer, D., Weisz, K., Mayer, C. and Kothe, G. (1991). *J. Chem. Phys.* **95**, 672.
Strey, R. (1992). Private communication.
Strey, R. (1993). *Ber. Bunsenges. Phys. Chem.* **97**, 742.
Strey, R., Schomäcker, R., Roux, D., Frederic, F. and Olsson, U. (1990). *J. Chem. Soc. Faraday Trans.* **86**, 2253.
Strey, R., Winkler, J. and Magid, L. (1991). *J. Phys. Chem.* **95**, 7502.
Strey, R., Jahn, W., Skouri, M., Porte, G., Marignan, J. and Olsson, U. (1992). *Structure and Dynamics of Strongly Interacting Colloids and Supramolecular Aggregates in Solution* (ed. S.-H. Chen, J. S. Huang and P. Tartaglia). Kluwer, Dordrecht.
Svetina, S. and Zeks, B. (1989). *Eur. Biophys. J.* **17**, 101.
Szleifer, I., Kramer, D., Ben-Shaul, A., Roux, D. and Gelbart, W. M. (1988). *Phys. Rev. Lett.* **60**, 1966.
Szleifer, I., Kramer, D., Ben-Shaul, A., Gelbart, W. M. and Safran, S. A. (1990). *J. Chem. Phys.* **92**, 6800.
Talmon, Y. and Prager, S. (1978). *J. Chem. Phys.* **69**, 2984.
Telo da Gama, M. M. (1987). *Mol. Phys.* **62**, 585.
Telo da Gama, M. M. and Gubbins, K. E. (1986). *Mol. Phys.* **59**, 227.
Telo da Gama, M. M. and Thurtell, J. H. (1986). *J. Chem. Soc. Faraday Trans. II* **82**, 1721.
Teubner, M. (1990). *J. Chem. Phys.* **92**, 4501.
Teubner, M. (1991). *Europhys. Lett.* **14**, 403.
Teubner, M. and Strey, R. (1987). *J. Chem. Phys.* **87**, 3195.
Tiddy, G. J. T. (1980). *Physics Rep.* **57**, 1.
Vinson, P. K., Sheehan, J. G., Miller, W. G., Scriven, L. E. and Davis, H. T. (1991). *J. Phys. Chem.* **95**, 2546.
Vonk, C. G., Billman, J. F. and Kaler, E. W. (1988). *J. Chem. Phys.* **88**, 3970.

Walker, J. S. and Vause, C. A. (1980). *Phys. Lett.* **79A**, 421.
Walker, J. S. and Vause, C. A. (1983). *J. Chem. Phys.* **79**, 2660.
Wang, Z. G. and Safran, S. A. (1990a). *Europhys. Lett.* **11**, 425.
Wang, Z. G. and Safran, S. A. (1990b). *J. Phys. France* **51**, 185.
Wang, Z. G. and Safran, S. A. (1991). *J. Chem. Phys.* **94**, 679.
Weeks, J. D. (1980). *Ordering in Strongly Fluctuating Condensed Matter Systems* (ed. T. Riste). Plenum Press, New York.
Wenzel, W., Ebner, C., Jayaprakash, C. and Pandit, R. (1989). *J. Phys. Condens. Matter* **1**, 4245.
Wheeler, J. C. and Widom, B. (1968). *J. Am. Chem. Soc.* **90**, 3064.
Widom, B. (1984). *J. Chem. Phys.* **81**, 1030.
Widom, B. (1986). *J. Chem. Phys.* **84**, 6943.
Widom, B. (1987). *Langmuir* **3**, 12.
Widom, B. (1989). *J. Chem. Phys.* **90**, 2437.
Wu, F. Y. (1982). *Rev. Mod. Phys.* **54**, 235.
Zhou, X.-L., Lee, L.-T., Chen, S.-H. and Strey, R. (1992). *Phys. Rev.* **A46**, 6479.
Zia, R. K. P. (1985). *Nucl. Phys.* **B251** [FS13], 676.
Zilker, A., Ziegler, M. and Sackmann, E. (1992). *Phys. Rev.* **A46**, 7998.
Zittartz, J. (1967). *Phys. Rev.* **154**, 524.

INDEX

Note–Page numbers in *italic* type refer to illustrations